LIVING DANGEROUSLY

Navigating the Risks of Everyday Life

JOHN F. ROSS

HELIX BOOKS

PERSEUS PUBLISHING
Cambridge, Massachusetts

ISBN 0-7382-0321-1
Library of Congress Catalog Card Number: 00-102396

The hardcover edition of this book was published in 1999 with the title "The Polar Bear Strategy."

Perseus Publishing is a member of the Perseus Books Group

Text design by Karen Savary Studio
Set in 11.5-point Electra by Pagesetters

1 2 3 4 5 6 7 8 9—03 02 01 00
First paperback printing, March 2000

Find us on the World Wide Web at
http://www.perseuspublishing.com

Dedicated to Diana Bosworth Ingraham

CONTENTS

ACKNOWLEDGMENTS *vii*

PREFACE: The Polar Bear Strategy *ix*

ONE: Maestros of Risk We're Not *1*

TWO: An Eccentric French Monk Launches a Revolution *11*

THREE: Even Lying in Bed Is Risky *35*

FOUR: Risk Management's Orwellian Side *53*

FIVE: Polar Bear Follies, or How Unconscious Risk Rules Cause Trouble *69*

SIX: I'm Risk Prone, You're Risk Averse *87*

SEVEN: War on the Soma *105*

EIGHT: Where Do We Draw the Line? *123*

NINE: The Myth of Zero Risk *147*

TEN: Our New Role as Risk Managers *167*

BIBLIOGRAPHY *177*

INDEX *187*

ACKNOWLEDGMENTS

The idea for this book on risk originated from an article I wrote for *Smithsonian* magazine that appeared in November 1995. At the magazine, I'd like to thank Don Moser, Jack Wiley, and Beth Py-Lieberman for their excellent work in helping me wrestle this difficult subject to the ground.

Special thanks goes to my agent, Theresa Park of Sanford J. Greenburger Associates, who realized that a book-length treatment existed long before I did—and continues to be a source of sage advice.

For a close reading of the manuscript I'd especially like to thank Baruch Fischhoff and Alexis Doster III.

In addition, I'm indebted to Phil Corfman, Mike Dole, Alex Doty, Judi Farkas, Timothy Foote, Deborah Grosvenor, James Knight, Phil and Jane Ross, John Starns, and John Thompson for their insightful comments and wonderful support. I also wish to thank the staff of the University of Maryland library system for their hard work in chasing down common and obscure references.

I'd like to thank those who have accompanied me on expeditions, especially Stark Biddle and Phil Brownell, who have provided so much grist for my mill.

Thanks, too, to the professionals at Perseus Books, who without exception proved dedicated, cheerful, and keen-eyed. Kudos especially to my editor, Amanda Cook, for her fine contributions and the care with which she ushered this project along.

And, finally, this book would not exist were it not for the tireless support and interest, sharp eyes and mind, and superhuman patience of my wife, Diana Ingraham.

PREFACE
THE POLAR BEAR STRATEGY

Danger from polar bears was far from our minds for the first few weeks of our two-month journey by canoe in the High Arctic. Our team of six men worried more about breaking through the ice-crusted water with our paddles and fending off clouds of mosquitoes when the wind died. Twin Otter float planes had dropped us into a lake deep in the midst of Victoria Island in Canada's Northwest Territory. Our position was northeast of Alaska and 300 miles north of the Arctic Circle, between ice cap and mainland, our goal to complete the most northerly, extended open boat trip on a white-water river in recorded history. We would paddle 215 miles, ending up where the Kuujjua River feeds into the Arctic Ocean.

Twenty-four hours of light and long days of paddling in the treeless tundra left us exhilarated but a little crabby with one another. An attempt to climb the tallest point of land in the Shaler Mountains, a rise of only 1,200 feet, ended in failure when a storm ripped our tents from their moorings. An epidemic of lost toenails resulted from our feet's constant immersion in the partially thawed permafrost, the consistency of granola in icy milk. So it was only when our tired crew neared the coast and saw footprints and a large hole in the ground, the result of a bear clawing to excavate a warren of small mammals, that talk turned exclusively to where this bear might be now. How recent were these diggings? With precious little concrete information about the local bear population, bear aggression, or other factors, each person used his own experience, fears, and beliefs to attempt to determine the severity of the risk we faced. The big question looming in front of us was what we would do if we came face to face with one of these creatures.

There was no question in our minds that polar bears are formidable predators. Among North America's largest land carnivores, polar bears are taller than grizzlies and can weigh more than 1,500 pounds. Their claws are needle sharp for traction on the ice and for subduing seals, their canines longer than those of grizzlies and ideally suited to tearing apart flesh. Polar bears are strong enough to pull a beluga whale out of the water, patient enough to wait hours above a seal's blowhole, and canny enough to hide their black nose with a paw as they stalk an unwitting seal. They're capable of killing humans and have done so often. Like sharks, they are attuned so perfectly to their environment that we humans can only be clumsy and ineffective interlopers in their world.

At dinner on the day we had seen the diggings in the morning, every team member aired his thoughts about how we should respond to this possible threat. By the time we had tipped our customary after-dinner drinks of rum and Kool-Aid, the discussion had degenerated to the level of terse declarations, accompanied by scowls and downcast eyes. One man grew close to hysterical, insisting that we take turns as sentry, all practice shooting the rifle we carried for protection, and no longer go to the bathroom alone. We brought a .30-06 rifle in part because no bush pilot would drop us off without some form of protection. Another team member ridiculed these ideas as "over the top." Should one person carry the gun? Where should we keep the ammunition? These questions soon grew so invested with emotion and vitriol that we couldn't even agree on whether to leave the rifle's safety catch on or off.

I'd like to be able to report that at that point I entered the conversation and articulated a sensible course of action in calm and deliberate tones, to which everyone else agreed forthwith. Certainly, out there on that featureless tundra, at least three days — or possibly a week — away from help, the danger of meeting a polar bear made me concerned. But I had no ready answer. The rum probably made matters worse. Although I definitely didn't support the plan of our most nervous member, I felt that some preparations should be made, and I was surprised at the level of disagreement in our group. Despite our similarities as white-collar professionals and our bonds as fellow explorers and friends forged out of many weeks cooperating with one another, our individual perceptions of the risk of polar bear attack remained vastly different.

The risks that each of us face in our day-to-day lives resemble that unseen polar bear that so preoccupied our expedition: danger signs are evident, but the actual hazards lie somewhere out of sight and have become difficult to quantify and address. Modern risks are vastly more complicated and far more difficult to understand than an angry bear, although they can be just as dangerous.

The possibility of an encounter with a polar bear, however, though clearly frightening, is at least tangible and concrete. But what about the risks of contracting an exotic cancer, for instance, a threat that's more difficult to put our hands around but could kill us just as surely? Hazards such as these fill our lives in modern times. Out on that tundra, because the six of us couldn't agree on the severity of the risk of meeting a bear, we neglected to come up with a strategy. Instead, paralyzed, we chose the worst possible course of action: to do nothing at all. We left the gun inaccessible in a duffel bag. In essence, we left our chances up to fate. On that expedition we happened to be lucky: we never did encounter a bear.

In large measure, we as individuals leave many of our day-to-day decisions regarding risk up to luck and fate in the same way. But I believe we can do better. The study of risk has blossomed lately into a far-reaching and fascinating science, from which both individuals and society can benefit enormously. There is always the chance of meeting that bear around the next bend of the river. We owe it to ourselves—and our families—to be as prepared as possible.

CHAPTER ONE

MAESTROS OF RISK WE'RE NOT

Let me say at the outset that I'm no stranger to risk. I'm not a daredevil, but for the last quarter century I have accompanied scientists to remote corners of the world, set up expeditions involving technical mountain climbing and white-water river running, and built a business in the precarious world of freelance writing. I've staked my life many times on my ability to assess the risks I face while climbing, paddling, or scuba diving. I must confess that the moments of greatest risk—on a mountain's knife edge, for instance—also provide instances of the greatest clarity and concentration. At these times, the presence of risk becomes tangible. Whereas fear socks you in the gut, the recognition that death is only a misstep away gives the experienced person a heightened sense of awareness, a constant and persistent pressure in the back of the neck. I know people who are addicted to those moments.

That said, imagine my surprise when risk sucker-punched me

in a doctor's office. I wasn't wearing crampons on my feet or snake chaps on my legs, nor was I out in the field or in any physical danger. The question of whether my pregnant wife should undergo an amniocentesis, an invasive procedure to perform genetic tests on the fetus, turned a pleasant conversation into one filled with concern and uncertainty. I felt that pressure on my neck—the kind I feel on a mountain ridge—but not the clarity that usually comes with it. The doctor spoke about the health of our unborn child, in which I had an all-consuming interest, yet I couldn't understand the significance of the number ratios she used to describe our choices. She might as well have been speaking Cantonese. Even with my considerable experience of the subject, I couldn't assess the risks presented to me. Nor could our doctor adequately interpret these risk choices for us. I glimpsed a world where my self-proclaimed expertise in risk didn't help at all. I didn't like it.

So I did what I'm trained to do: research. I read extensively and spoke with people about what had happened in the doctor's office, trying to make sense of it so that we could make the right decision. What I learned astounded me—a whole new science of risk analysis exists, and mountains of technical literature explore aspects of risk that I had never imagined. Pushing beyond the technical jargon, I felt like an explorer uncovering a new world. Before I knew it, eight years had passed and I'd published two major magazine articles on the subject and sold a book idea on the topic.

Questions of risk permeate every waking second of our lives. But as a large number of recent risk studies reveal, most Americans remain woefully ignorant and misinformed about even the most basic issues concerning risk. Hopelessly so, in fact.

My getting caught off guard in the doctor's office is a good example of the condition. Teddy Roosevelt once exhorted, "Far better is it to dare mighty things, than to take rank with those poor spirits who neither enjoy much nor suffer much." He pretty much summed up the American attitude toward risk: dealing with risk is just a matter of a steely will and resolve, mixed with a good shot of courage and fortitude, like soldiers taking a hill by putting their heads down and charging. America is a culture of risk taking, and Americans are committed to the idea that taking risks leads to wealth and happiness—and, furthermore, that we're good at it. "It is only by risking our person from one hour to another," said William James, "that we live at all." Americans applaud the entrepreneur who risks everything, the young film student who maxes out his credit cards to make his first movie, the test pilot who steps into an experimental aircraft. After all, this is the nation of immigrants who gave up everything to come here. Conversance with risk, it seems, is an American birthright.

However, our conception of ourselves as risk masters *par excellence* breeds an arrogance that prevents us from realizing that the landscape of risk has changed radically during the last several decades. Our ancestors never faced making decisions about whether to have an operation or a course of treatment based on the statistical probability of its success. They never had hard and fast data proving that smoking, a diet high in fat, stress, high blood pressure, and elevated cholesterol can cause specific health problems thirty years in the future. Our grandparents didn't have the choice that my wife and I had about whether to risk having this genetic test carried out on our baby in the womb. They didn't need to consider writing a living will that specifies whether a ventilator should be used to prolong their life. And they never faced what to do with the prodigious quantities of data that falls into our laps every day, the fruits of the information revolution. Not only is the information an average person receives more complex and voluminous than ever before, but the types of risks he or she faces are changing like quicksand under our feet. Diseases that stalked our grandparents and great grandparents have been banished, while a battery of new evils, as diverse as antibiotic-resistant strains of bacteria and bioterrorism, have taken their place.

As a result, we're continually confronted both with new orders of risks and revelations that substances and products once thought safe are in fact not so. In the 1950s, panics in the health arena were rare: traces of carcinogenic weedkiller in cranberries from the Northwest, fluoride in water, strontium 90 in milk, fears of polio epidemics. Today, newspapers print screaming headlines about artificial hormones in our water supply and shrinking sperm counts, television news bulletins blare warnings about new outbreaks of *E. coli* poisoning and the emergence of new bacterial strains resistant to antibiotics, magazines sound the alarm about the possibilities of an asteroid colliding with Earth and global warming altering the weather. Little understood "syndromes" abound: sick building, toxic shock, sudden infant death, chronic fatigue.

We have begun to encounter ethical dilemmas of which we have no previous experience. Many risk situations reach beyond our experience and are not biologically intuitive. In modern America, many of the risks we face lurk unseen just around the corner, like that threat of polar bears that I described in the preface to this book. Our inability to agree on how to prepare against a bear attack threw our expedition into a tailspin. When should we take a risk seriously enough to change our behavior? Every day we must decide how this information should affect our lives. Journalists, bureaucrats, scientists, and regulators often leave information about risks vague and ill defined, letting the public wrestle with their fears and emotions.

Despite these changes in the fabric of modern life, most Americans still evaluate risks in the same ways as their grandparents and those before them did, using antiquated logic, misguided assumptions, and emotion. For most of history, humans have relied on horse sense, anecdote, past experience, and intuition to get through life's neverending parade of decisions. We listen to folk wisdom, repeat what others have told us, and frequently react emotionally. Some of us rationalize not wearing seat belts on the basis of knowing about one person who survived a terrible wreck only because he *wasn't* wearing a seat belt and was therefore thrown clear. It doesn't matter that numerous studies have shown that a person who wears a seat belt has a 42 percent better chance of surviving an accident than one who does not. Using the same kind of emotional logic, a woman may choose not to have a mammogram because of her fears about x-rays, for instance, although the possibility of cancer is a far worse risk than the radiation exposure from an x-ray. "Most people think dramatically," noted the famous American jurist Oliver Wendell Holmes, "not quantitatively." It's certainly not true that Americans are stupid or foolish — it's simple ignorance that's at play here.

That ignorance, however, gets us into trouble. Time and again, Americans worry about risks that are insignificant while virtually ignoring truly substantial threats. The same people who succumb to the dangerous behavior known as road rage also pay a lot of money for free-range chickens and organic produce because they fear eating minute quantities of animal growth hormones and antibiotics. We spend a billion dollars on tamperproof packaging when seven people die from poisoned Tylenol tablets, yet we do nothing about the fifty children who drown every year in 5-gallon buckets. Some people avoid ocean swimming because they fear sharks will attack them, but go bungee jumping instead. We smoke billions of cigarettes, yet think about banning artificial sweeteners because they might raise our risk of cancer infinitesimally. We panic about mad cow disease, botulism, and flesh-eating bacteria, yet three-quarters of us don't eat the recommended five daily servings of vegetables and fruits, the foundation of good health. Old habits die hard, but die they must. Misunderstandings and misjudgments about basic risks cost us billions of dollars every year, take thousands of lives, and cause pandemics of fear. Charging the hill no longer works as a strategy for dealing with risk; mastering the complexities of risk today involves far more finesse and subtlety.

Even the nuances that the word *risk* has taken on lately reveal how deeply enmeshed it has become in the American psyche and vernacular. At a checkup, a doctor informs you that high blood pressure is a "risk factor" for heart disease or discusses the "risks" of a regimen of postmenopausal estrogen. Newspaper headlines warn that inner-city youths are "at risk," a radio report talks about the financial "risk" to your stock portfolio if a certain foreign market collapses, and a televison news program reports on the "risk" that a devastating hurricane will hit the coast. Hundreds of times over the course of a normal person's day, the word risk is invoked. "What's the risk?" seems like some nationwide mantra. Quite simply, "risk" — whether used to describe physical, financial, social, or ethical situations — has become one of the most important words in the modern American lexicon.

While our understanding of the nature and dimensions of risk has grown astronomically, the definition of risk remains quite simple: the possibility of loss or harm. An activity, condition, or substance is risky if it could result in your losing something — life, money, morality, beliefs, health — or harm a life or material object. "Risk," states a recent report on risk analysis by the National Academy of Sciences, "is a combination of the probability of an event — usually an adverse event — and the nature and severity of the event." Of course, dealing with risks and hazards is nothing new in human society. What's changed of late — to a revolutionary extent — is the way in which scientists can now understand, quantify, and identify risks. The birth of the multidisciplinary science of risk analysis in the 1970s has fostered journals, university departments, numerous careers, and a growing pile of influential papers. But the foundation of this new science in fact lies in a remarkable revolution in Western civilization that occurred four centuries ago.

It started with the musings of an eccentric and brilliant seventeenth-century French monk named Blaise Pascal, who proposed a new and rigorous approach to thinking about how future events might unfold called probability theory. In essence, although on a limited scale at first, probability theory enabled its practitioners to quantify the odds of two different events occurring and then compare them. The effect of this simple but remarkable work was like letting a powerful genie out of its bottle. The insights gained by science and technology as a result, along with the new tools developed for analyzing risk decisions, soon radically changed the way humans thought about uncertainty and regarded the future. The theory bore directly on how people make decisions and, consequently, how they live their lives, even among people who don't know the first thing about statistics. Both as individuals and as a society, Americans and America still wrestle with the consequences of Pascal's work every day of their lives.

Meanwhile, new and evolving ideas and technologies in risk analysis continue to turn old notions on their head and raise vexing questions. Take a deep breath and hold it. In your lungs right now lurk molecules of nature's most deadly toxins: dioxin, mercury, arsenic. Quantitative techniques in toxicology can now isolate poisons in amounts of one in a quadrillion, the equivalent of a shot glass of vermouth mixed with enough gin to make a martini the size of the Great Lakes. Science now understands that toxins—and their attendant risks—are everywhere. Even the bones of our loved ones release radiation that strikes our bodies. What is safe and natural? Are there thresholds below which a toxin is safe?

New insights on the cellular level reveal the prodigious risks faced by every cell in our body, each of which is blasted on average once every ten seconds by radiation or destructive chemical agents. Our new understanding of risks to cells has stimulated new and intriguing theories on cancer and aging. Science and technology in the latter part of this century alone have introduced and identified a new type of human-made risk: low-probability events with catastrophic impacts, such as a nuclear power plant meltdown. Engineers can use sophisticated modeling and probability theory to tease out the nature of such risks stretching far into the future. There's global warming, the impact of artificial hormones dumped into the environment, holes in the ozone layer, and the possibility that a large comet or asteroid will strike the Earth. How much is the nation willing to pay, how much will it give up now, to forestall something that might happen well in the future, perhaps long after all of us have died? These questions and issues won't go away, and as they continue to accumulate, they force us to make more and more complex decisions — or to ignore them completely.

As I delved deeper into the subject of risk — the piles of statistics on risk, ideas about how Americans perceive risk, how risk frames difficult issues — it struck me that I was learning a new language. A person can try to work with new risks revealed in modern life by memorizing and hoarding information. However, that's akin to trying to learn a foreign language by just memorizing vocabulary. We are soon awash in words that we don't know how to string together. Fluency in a foreign language involves knowledge not only of words but also of grammar, syntax, even the culture behind the language, including an idea of its history and myths. Increasingly, we look at the world through the lens of risk. Therefore, to make sensible decisions and operate independently of those who might manipulate us, we must become fluent in the language of risk. The two critical elements of this language are (1) demanding accurate and usable risk information from experts and (2) knowing how to begin to interpret that information.

It's not easy. I've had people at dinner parties rise up out of their chairs shouting when a certain risk-related subject came up. I've found that discussing risk is often like touching a raw nerve because it concerns what we most fear to lose: be it a breast, a prostate, a pristine environment, civil liberties, or a loved one. Limited resources force both individuals and society to make tough and often controversial decisions. That makes learning the language of risk more important than ever before. An aspect of understanding the language and theory of risk analysis revolves around the recognition that different people have different tolerances for risk. Why are some people bigger risk takers than others? As psychologists push deeper into the evolutionary, genetic, and personality roots of risk-taking behavior, we may be able to learn where individuals fall on the risk-prone/risk-averse scale—and to understand how our biological and psychological makeup influences decision making.

Ultimately, learning the language of risk confers power. If Americans don't become fluent in the vocabulary and ideas of risk analysis, they won't be able to take advantage of the remarkable new insights that science routinely brings forth. Furthermore, they won't be able to understand what the experts are telling them. I felt that way in the doctor's office. If we fail to understand basic risk issues, the experts will happily make decisions for us and point to our ignorance as a justification. Americans will cede away the most valuable commodity in a democracy: the right of the people to make their own decisions. That is certainly not a risk that you or I should be willing to take.

CHAPTER TWO

AN ECCENTRIC FRENCH MONK LAUNCHES A REVOLUTION

By all accounts, the seventeenth-century French mathematician and monk Blaise Pascal was a rather miserable fellow, prone to stomach disorders, bouts of insomnia, and mental illness. Worse still, he spent much of his adult life torn between the pleasures of the earthly world (to him, an enjoyment of mathematics and science) and the ultrastrict world of the Jansenists, a fanatical sect devoted to banishing all but rigid adherence to religious matters. An indication of his conflicted soul comes anecdotally near the end of his tragically short life. A toothache prevented him from sleeping for days on end. To take his mind off the pain, he reputedly thought about the cycloid, a curving line generated by a point on the circumference of a circle as it rolls along a straight line. Miraculously, thinking about the cycloid deadened the relentless throbbing. The disappearance of the pain, figured Pascal, was an indication by God that it was okay this once to think about the cycloid and not dwell only on matters of faith. He spent the next eight days thinking about cycloids. However, the pull of religion in his life eventually won out and he would quit mathematics altogether, a tragic loss lamented by future generations of mathematicians. But before this happened, Pascal would confront one of humanity's thorniest and most difficult conundrums with remarkable success.

The question of the existence of God has challenged the best minds for thousands of years. Pascal decided to look at this question obliquely, examining and then comparing the consequences of believing or not believing in God. "Pascal's Wager" became an elegant, fresh, and novel approach to the question of belief in God. While it may appear subtle — even semantic — today, this new way of thinking — eventually known as probability theory — was to revolutionize the way in which everyone from Nobel-winning physicists to average Americans would look at the world around them. Every corner of modern life is influenced by the ideas first set forth by this brilliant French monk. Pascal's work lay the groundwork for thinkers who later pioneered discoveries more stunning than those made by the explorers who had sailed the high seas: the discovery of probability theory, statistics, sampling, and the means to bend numbers into powerful tools to analyze the possibility of events occurring in the future. For the first time, a scientific examination of uncertainty became possible.

Today, we tend to associate probability and statistics with arcane and dusty corners of science and economic theory. This obscures the fact that probability and statistics have exerted profound changes in the way individuals, in a personal and immediate way, regard the world, think about risk, and visualize the future. Humans still and will always face loss or harm in their lives, but Pascal and later thinkers focused on calculating precisely the possibility of that loss or harm occurring, the realm of the uncertain future. By better understanding the likelihood of future events happening, Pascal gave scientists a powerful tool to examine the shape of the future itself, the closest thing to looking into a crystal ball that humankind is ever likely to have. While revolutions in science, industry, and political institutions have long drawn the attention of historians, the revolution wrought by probability theory on western thinking has been neglected. It's a truly remarkable story.

Pascal was cut from a different cloth than most of us. As a boy, sickness kept him from school in Paris, but his genius didn't take long to bloom. He mastered Euclid's *Elements* by the age of thirteen, and by fourteen he was attending weekly meetings of France's top mathematicians. As a teenager, he authored elegant proofs and wrote an essay on conics, the study of the curvature of cones. At eighteen, to lighten his father's workload as a tax collector, he invented the world's first mechanical calculating machine. Pascal's insights into probability, however, were not triggered until later, when a man brought him a puzzle.

Chevalier de Méré was an arrogant nobleman who dabbled in mathematics, mostly putting his hunches to use in the gambling parlors of Paris. It was his love of games of chance that prompted him to bring a centuries-old brain teaser to some of the leading mathematicians in France, among whom was Blaise Pascal. In fact, an Italian monk, Luca di Paciuolio, had put forth this same question two hundred years before, and no one had been able to solve it.

The "problem of the points" involves two players who are in the midst of a *balla* game, winner take all. One is winning. Unfortunately, the match is broken off before it's finished. How does one equitably split the prize money? Pascal's ever-curious mind fixed on this problem. He dismissed the solution calling for the even split of the purse, because it did not adequately reward the person who was ahead. Nor would giving all the money to the person ahead adequately acknowledge the possibility that the losing person could come from behind and win should the series be completed. Pascal realized that the answer lay in determining the odds of each player winning. Inherent in this simple notion was the very idea of probability—establishing the numeric odds in favor of or against an event occurring in the future.

For consultation on this thorny problem, Pascal turned to Pierre de Fermat, a lawyer in Toulouse who studied mathematics on the side. He independently discovered analytical geometry, and some mathematicians call him the greatest mathematical mind of all time. Pascal wrote to Fermat, and the two men corresponded enthusiastically throughout 1654, wrestling with the problem of the equitable division of spoils in the unfinished *balla* game. In August of that year, Pascal adopted what's become known as Pascal's triangle, a table of numbers shaped in a triangle, cleverly and graphically illustrating the odds of a certain event occurring—be it the outcome of a game of balla, dice, or other game of chance. The beauty of Pascal's achievement lay in its recognition that mathematical principle, not just the bettor's hunch, can be applied to figuring out the odds of an outcome in a game of chance. In short, Pascal and Fermat had established the theory of probability: the idea that while one could never be sure what would happen in a game, one could calculate the odds of various outcomes with consistent, mathematical precision. They thus concluded that to be divided equitably, the spoils of the balla game should be split according to the odds each player had of winning the series at the point where the games were interrupted.

Although no one before Pascal had formally described probability, others before him had wrestled with the issue, as is evident from sections of the Talmud and sections of ancient Arabic texts. Humankind's long fascination with gambling certainly provided an incentive. Figuring out the mathematical odds in a dice game or hand of cards gives a gambler remarkable advantages. Indeed, the Greeks and Romans — inveterate gamblers that they were — nearly came up with a formalized theory of probability theory; historians of science argue about which factor most stymied their efforts — their cumbersome numeric system, a deterministic world view, or the strong elements of piety in their societies. One historian remarked, tongue in cheek, that in classical times a person with a rudimentary understanding of probability could have gambled so effectively that he could have "won himself the whole of Gaul in a week." While gamblers of all eras have picked up much about basic probabilities from experience and intuition, the power of Pascal's new theory took understanding of odds a large step forward.

The year 1654 proved pivotal for Pascal. Despite what must have been a scintillating correspondence with Fermat, Pascal increasingly felt religion calling him away from these worldly pursuits. His poor health, bouts of depression, and a series of disappointments found him increasingly frustrated with secular life. An event late in 1654 made up his mind. One day at Neuilly, his team of horses bolted while crossing a bridge. The force and suddenness of the movement broke the trace, and the horses plunged over the edge of the bridge, while a breathless Pascal sat immobile on his now horseless carriage, which remained atop the bridge. The frail genius interpreted his miraculous escape from death as divine intervention. He subsequently sold his worldly possessions and, at thirty-one years of age, moved to a monastery and took up the Jansenist cause with passion.

Before he died eight years later, Pascal went on to publish important religious treatises. But it is two pieces of paper, each covered front and back with handwriting scribbled in all directions, that draws most attention today. He took the same principles of probable outcome he had worked out with Fermat about the balla game and applied them to an entirely different matter. He asked: "God is, or he is not. Which way should we incline?" Although Pascal intended his writings as a defense of faith and Christianity, his musings on these two pages proved that his nascent ideas of probability theory could apply not just to games but to questions that span all of human experience.

Reason, Pascal understood, could never establish definitively whether or not God actually exists. He did realize, however, that he could bring logic to examine the consequences of believing or not in the same way as he had examined games of chance. Pascal believed that just as games of chance had only a certain number of outcomes, so did the act of belief or disbelief. Could an examination of the odds in this case influence his behavior? For Pascal, belief in God could have two outcomes, depending on whether it was justified or not. If God didn't exist, nothing would happen to the believer. If God did exist, however, the believer would be blessed with eternal salvation. The nonbeliever also faced two possible outcomes. If God didn't exist, he would suffer no consequences. If God did exist, though, the nonbeliever would face eternal damnation. The nonbeliever faced either no consequences or hell; the believer no consequences or heaven. Faced thus with a choice between heaven or hell, Pascal figured, it made sense to open oneself to faith. Many people today adopt Pascal's reasoning when thinking about God. How many religious fence-sitters actually fear the consequences of not believing so elegantly demonstrated by a sickly genius nearly 350 years ago?

While later philosophers have criticized the wager for its cynicism, Pascal's logic remains solid. Historian Ian Hacking writes that the logic gave birth to decision theory, "the theory of deciding what to do when it is uncertain what will happen." According to Hacking, Pascal not only was the first to figure out the theory of probability, he was also the first to take the grand leap of applying the logic to situations outside gaming. Here is the very heart of modern risk analysis and management: how to evaluate the consequences of individual and group actions, measure them, and consequently make better decisions. By using the power of numbers, probability, and later statistics, humankind mastered the use of powerful tools to harness the information of the past, project these findings into educated guesses about the nature of future, and flesh out the likelihood of uncertain events occurring. It became possible for the first time to analyze risk rigorously and scientifically. Pascal's logic and the fledgling science of probability found ready applications in the legal work of Gottfried Wilhelm Leibniz, the development of actuarial tables and the emergence of commercial insurance at Lloyd's Coffee House in London, and the field of medicine. Today, elements of probability are found in every scientific discipline and experiment. Where there is probability, there is risk analysis.

The word *risk* entered the English lexicon at about the same time as Pascal worked with probability theory during the mid-seventeenth century. Not once did English literature's master wordsmith, William Shakespeare, who died in 1616, use the word in his plays or poetry. Certainly, the Bard understood the concept behind the word used today — the idea that an activity or event could result in the possibility of loss or harm. Of course, Shakespeare, Alexander the Great, or a medieval knight preparing for a joust all understood perfectly well the consequences of their actions, regardless of whether or not they used a word to describe it. No one needed to understand probability theory to know that different courses of action could change the outcome of a dangerous situation. Shakespeare used the word *hazard* more than three dozen times. "Men that hazard all," says Morocco to Portia in *The Merchant of Venice*, "Do it in hope of fair advantages." The verb *hazard*, which means to venture, is derived from an Arabic dice game *az-zahr*. Taking the other meaning of *hazard*, a source of danger, the difference between *hazard* and *risk* may seem slight or semantic. However, the distinction between the two reveals how significantly our world view has changed since Shakespeare's time, largely due to the theory of probability. *Hazard* refers to an activity or technology that poses a threat to humans and what they value. In contrast, the more abstract concept of risk centers on the possibility — or probability — of that hazard or danger being realized. Within the province of possibility and uncertainty lie some of the most vexing and involving questions facing human beings.

The word *risk* appears in English in the 1660s, probably adapted from the Italian term *risico*, which means hazard or peril. Etymologists agree that the modern word *risk* derives from this early Italian word but can't place its precise origin. Considering that Italy dominated Mediterranean trading from the Crusades until the discovery of America, it's likely that the term originated there. The Italians acted as the financiers of the Catholic church; they developed double-entry bookkeeping, the draft bill of exchange, and other key elements of business practice. A Spanish sailor on board his galleon used the term *risco* to describe a steep, abrupt rock—every sailor's nightmare. The Spanish *arriesgar* means to venture into danger or, literally, "to go against a rock." It's not surprising that marine ventures spawned the earliest examples of insurance and inroads into risk analysis. For most of recorded history, extended ship voyages consistently represented the most daring forays into the unknown. Risk analysis proved a critical element in a ship captain's and ship owner's survival, both physically and financially.

Pascal's complex insights into probability were popularized by a Dutchman named Christiaan Huygens in a treatise on gambling, *De Ratiociniis in Ludo Aleae*, published in 1657. "You will find here," wrote John Arbuthnot in the preface of his English-language translation, "a very plain and easie Method of the Calculation of the Hazards of Game, which a man may understand, without knowing the Quadratures of Curves, the Doctrin of Series's, or the Laws of Centripetation of Bodies, or the Periods of the Satellites of Jupiter. . . . There is nothing required for the comprehending the whole, but common Sense and practical Arithmetick." Probability theory brought with it not only a handful of mathematical principles but a new way of thinking about the future. Analysis could reveal the chances of certain outcomes. Toward the end of his preface, the mathematician Arbuthnot notes, "The Reader may here observe the Force of Numbers, which can be successfully applied, even to those things, which one would imagine are subject to no

Rules. There are very few things which we know which are not capable of being reduc'd to a Mathematical Reasoning, and when they cannot, it's a sign of our Knowledg of them is very small and confused. . . ."

Lord William Thomson Kelvin would later sum up the same sentiment in his famous dictum: "when you can measure what you are speaking about, and express it in numbers, you know something about it; but when you cannot measure it, when you cannot express it in numbers, your knowledge is of a meagre and unsatisfactory kind."

While numbers provide a universal language of expressing values, probability theory provided a way to describe events — and risks — using numbers. The ability to describe the possibility of some future event with numbers — a percentage, for instance — led to something remarkable: it became possible to compare accurately the severity or frequency of one risk with that of another. If the expression of risk could be defined in the universal language of numbers, then a person could compare the chances of getting killed in a shark attack with those of being struck by lightning. In fact, just about any potential risk could be compared with another if enough data were available. This spawned the simple but revolutionary idea of relative risk. When this idea came into full force late in the twentieth century, it began subtly and not so subtly reordering our priorities and forcing us to reexamine the risks we faced in new ways.

One of the first proponents of relative risk was an English haberdasher named John Graunt. In the late seventeenth century, Graunt came to the idea of relative risk by understanding the importance of gathering data as a tool to divine patterns in past events that might shed light on the nature of future occurrences. In doing so, Graunt became one of the world's first number crunchers. Since William the Conqueror had the *Domesday Book* compiled as a survey of the English population and property in 1086–1087 AD, the English have sporadically assembled data and numbers. Church parishes often kept count of the number of births and deaths at the whim of the current parson. For centuries, however, no one derived conclusions from these figures. No one chose openly to ponder the significance of the number of sick, the number of girls versus boys, or any other comparative information that today seems to invite analysis. Perhaps such observations would have seemed to question the authority of God. In the early seventeenth century, such demographic data became a source of great interest after bubonic plague had reemerged with devastating force in Europe. In London, the Privy Council asked the mayor to supply a tally of plague deaths. By the end of the century, the "Bills of Mortality" appeared weekly, listing the number of births as well as deaths and their causes. The plague came in waves for decades, some outbreaks mild, others severe. Reputedly, the rich residents of London used these numbers as a barometer of its virulence, fleeing the city for sanctuary in the country when deaths rose above a certain number.

The Bills of Mortality was published throughout the seventeenth century, and the documents that survive provide an intriguing glimpse into who the grim reaper hauled away. They also reveal the nature of medicine at that time. The bill summing up the year 1665 in London includes 65 categories, among them the following: "Frighted—23; Found dead in the streets, fields, and &c.—20; childbed—625; Collick and Winde—134; Stopping of the Stomach—332; Grief—46; Aged—1545." But plague dwarfs all other entries with a frightening 68,596. Graunt perceived greater value in these numbers than a mere indication of the virulence of the plague at any one time. He gathered the death tables between 1604 and 1661 and analyzed them, publishing his results in the "Natural and Political Observations Made upon the Bills of Mortality." A later demographer would liken John Graunt to Columbus for this groundbreaking work.

Graunt's remarkably modern and prescient words gave voice

to the idea of relative risk: "Whereas many persons live in great fear and apprehension of some of the more formidable and notorious diseases following; I shall only set down how many died of each: that the respective numbers, being compared with the total 229,250 [deaths over 20 years], those persons may the better understand the hazard they are in." He made a fascinating discovery: all people worry about their own death, but most people are afraid of dying from causes that are actually not very likely to kill them. While the plague proved a notable exception, he noticed that people worried about splashy, dreadful deaths, not the more mundane causes of mortality that would most likely claim them.

Graunt's insight was extraordinary. He looked at the mortality figures in a new way: not anecdotally, but as a database of information from which to seek out new pieces of information. It was like finding the wreck of a treasure ship. He came up with the first reasonable estimate of the population of London, the ratio of males to females, and the emigration of people to the countryside during the plague. He even overturned a popular notion that London's population explosion was due to the special fecundity of Londoners; the numbers revealed that immigration was responsible. His data also suggested that the population of London took only two years to rebound after even the worst plague years.

Perhaps most remarkably, Graunt conjectured about the cause of the plague itself. It's easy to forget that despite some important medical breakthroughs, they had no clue as to the origin of the plague at that time. The exact cause of the plague, bacteria traveling to humans via fleas carried by black rats, would not be discovered until 1894 as a result of the work of French bacteriologist Alexandre Yersin. In Graunt's day, "miasmas," literally, heavy vapors from the atmosphere topped the list of suspected culprits. Graunt noted that the number of plague deaths seesawed abruptly from one week to another. On the basis of this pattern, he surmised, "The contagion of the plagues depends more upon the disposition of the air than upon the effluvia from the bodies of men." In the same way that Pascal didn't need to know whether God did or did not exist to think about the consequences of believing or not believing, Graunt didn't need to know what caused the plague to draw inferences about the nature of its transmission — in this case, the fact that the plague was generally not transmitted from person to person as small-pox was. Graunt showed that data, through inference, could point toward causality. He showed that one could crunch data from the past and use it to think about the future in new ways.

Eventually, Graunt's pioneering work evolved into the science of epidemiology, in which careful studies of patterns of sickness over populations revealed links between sickness and certain activities, diets, and lifestyles well before science understood the mechanisms of transmission or causes. The collection of vital statistics in the twentieth century has proved a major factor in the ability of science to increase longevity and reduce risks to our health.

Graunt also attempted to estimate the longevity of the average seventeenth-century Englishman. Although some of his logic proved flawed, he laid the groundwork for the actuarial tables that are the foundation of the modern insurance industry. Without a sense of how long the average person lives, based on careful analysis of data, the insurance industry would have no benchmarks upon which to base its premiums and analyze its risks. In the decades and centuries following Graunt, Pascal, and Huygens, a cast of others added more sophisticated tools to the growing arsenal with which science examined the nature of risk with ever-increasing rigor, clarity, and precision. Each new tool added another beam of light that illuminated more about the shape and disposition of events that had yet to occur.

Mathematicians after Graunt became fascinated not only with what averages might reveal, but about what nonaverage pieces of information could tell them. In other words, if the average American male life expectancy is seventy-three years, how normal or abnormal is it if a man dies at seventy-five or seventy? A Frenchman named Abraham de Moivre, who lived between 1667 and 1754, became intrigued with the well-known observation that variations exist among a similar set of phenomena or populations. Take the height of all Americans, for instance, or the loudness of warblers' songs, or the circumference of mature pine trees. Differences naturally occur in each grouping. Some Americans are quite short, others are tall, but most are near average height. De Moivre discovered that the distribution of these variations often follows a particular curve, which looks something like the shape of a bell when graphed onto a piece of paper. Regardless of the unit—height, song volume, tree circumference—the array of variations frequently takes the same shape. His discovery would become known as the normal distribution or, more popularly, the bell curve.

De Moivre was a man without a country. Forced to choose between fleeing his native France or giving up his faith, the brilliant mathematician sought refuge in England. Despite his brilliance, his French background prevented him from ever obtaining a chair in mathematics. He eked out a living as a traveling teacher of mathematics; later, he set up shop at a table in Slaughter's Coffee House in Long Acre and dispensed advice to gamblers and underwriters. His close friend, Isaac Newton, would collect him every evening and bring him home for conversation.

To visualize de Moivre's bell curve, imagine a pile of blocks on a basketball court. Each block represents a single person in the surrounding community. Written on each block is the person's height in feet and inches. Arrange the blocks in a line across the floor from one basket to the other, beginning with the block representing the shortest individual and ending with the block representing the tallest person. Pile blocks representing the same heights atop one another (i.e., all blocks that read 5 feet 6 inches would be stacked on top of one another). If the blocks are viewed from the side (on the bleachers), the shape will resemble the profile of the Liberty

Bell. Do this in Japan or among the Masai and the shape will be the same. Despite the fact that the heights may be different—the tallest Japanese may be shorter than the shortest Masai—the variation within each group will be the same, resulting in the same bell-shaped form. Most people's heights cluster around the average, or the middle of the basketball court. That's the crown of the bell. The broad, gently sloping shoulders of the bell, moving from the center of the court toward each net, represent the people who are slightly taller or slightly shorter than the average. They still represent quite a few people, who vary from the average only by several inches. The sides of the bell then slope precipitously downward to the handful of blocks at either extreme that represent the very tall and the very short. Standardized testing often reveals a bell curve in student abilities—most students cluster around the average, while less and more gifted students are fewer in number. One of the most familiar uses of the normal distribution is when a professor decides to throw out traditional grading and grades "on a curve." The teacher simply finds the average score and awards those students a B or B–. The rest of the students receive a grade based on how they did compared with the average.

If enough samples are taken, the normal distribution describes many naturally occurring populations in nature. De Moivre mathematically described the essence of this curve, so that the units of measurement (height in inches, chirps, or whatever) did not matter. His work grasped the essence of the process known today as "sampling," the idea of determining patterns in a large population by examining only a few individuals. Shoppers intuitively sample at the grocery store. In selecting green beans for dinner, for instance, the typical shopper examines only a few to determine the overall quality of the lot. Satisfied that the few good beans scrutinized represent the overall quality of the bin, the shopper then scoops up handfuls without further close examination, knowing that only a few beans in the bag will differ much from the acceptable average. De Moivre introduced a way to flip this intuitive sense of sampling and look at it from another angle. Scientists can use his calculations to determine whether or not a sample number of observations is representative of the larger population by examining how closely they match the normal curve. The technique proves invaluable to scientists, demographers, physicians, and political pollsters who can't possibly measure all examples of what they're studying. By sampling a small group and statistically examining the variation, scientists can make a good educated guess about what the entire population looks like. That's why political consultants can take exit polls during an election and enable the network news to give an accurate assessment of who won each political race long before all the votes are tallied. Today, standard deviation is a critical element of all scientific endeavors to determine whether data are accurate or contain errors. On the flip side, understanding the extent of natural variations among a particular subject enables an investigator to make predictions about the future. An insurance company, for instance, can look back over several years of data on the number of residential fires, make some statistical calculations, and then predict the number of fires within a range that will occur next year. If the actual number of fires turns out to be much higher than expected, unusual circumstances may be responsible: an arsonist or a new brand of faulty space heaters hitting the market, for instance.

A rather peculiar English genius named Francis Galton added further innovations to the concept of the normal distribution in the late nineteenth century, fine-tuning our ability to forecast the future with a greater degree of accuracy than ever before—and thus enabling us to analyze risk with greater precision. A man of uncommon talents, Galton worked in a variety of fields, performing pioneering work in fingerprinting as well as describing anticyclones in weather for the first time. But his real contribution was in statistics. His hagiographer and a noted mathematician in his own right, Karl Pearson, wrote that Galton "modified our philosophy of science and even of life itself." For some, Galton's legacy appears darker. He also created the science of "eugenics," the idea of manipulating human populations to bring out the "best" traits and thus direct their own evolutionary future. (A particularly nasty interpretation of eugenics would be adopted during the next century when the Nazis in Germany attempted to create a perfect so-called Aryan state by systematically killing off Jews, gypsies, homosexuals, and men with long criminal records.)

Born in 1822, Galton inherited a comfortable living from his father and never had to work. He therefore had time to satisfy his peculiar insatiable curiosity: he liked to measure everything he came across. On a trip to southwest Africa in 1849, he came upon a full-figured Hottentot woman. "I profess to be a scientific man, and was exceedingly anxious to obtain accurate measurements of her shape," he notes in his journal. Unfazed by not speaking the language, Galton picked up his sextant. "I took a series of observations upon her figure in every direction, up and down, crossways, diagonally, and so forth . . . I boldly pulled out my measuring tape, and measured the distance from where I was to the place where she stood, and having thus obtained both base and angles, I worked out the results by trigonometry and logarithms."

His interest in measuring began to take an even more interesting form when he dovetailed it with an interest in evolution. Like many intellectuals of the mid-nineteenth century, Galton had become fascinated with the work of Charles Darwin, who coincidentally was Galton's first cousin. Darwin had recently published *On the Origin of Species* and introduced the world to the revolutionary idea of natural selection in nature. Galton became curious about the passage of traits from one generation to another. After a discussion with Darwin, Galton set up an experiment. He convinced nine friends, scattered across England, to help him by planting sweet pea seeds for him in their household gardens. Along with the eight others, Charles Darwin did his share, planting seventy seeds, each carefully presorted by weight and diameter. The participants received exact instructions for planting, upkeep, and harvesting.

When Galton collected the harvest, he carefully sorted the seeds of the offspring and measured them. Then he compared the size and weight of the parents and offspring. He found a curious thing. The seeds of the offspring were less extreme in size than their parents. Instead of creating even larger seeds, the largest parental seeds most often produced smaller offspring. The smallest parental seeds, conversely, tended to produce larger seeds. While the sizes of the offspring could be graphed and showed a normal curve, the lips of the bell didn't encompass as broad a range of sizes as the bell describing the sizes of the parents.

Galton pondered the meaning of this information, searching for a pattern. What he found, and would later confirm in human studies, was that the ancestry of an organism—the long line of individuals going way back in time—exerted a profound genetic pull on the offspring. He called this idea the law of "reversion to the mean," which later became known as "regression to the mean." Residing in our genes is a deep memory of what has come before us. This vast silent legacy strongly influences human traits, such as height and intelligence. In essence, they pull these traits inexorably back toward the average or mean. He wrote: "The law is even-handed; it levies the same heavy successional-tax on the transmission of badness as well as of goodness. If it discourages the extravagant expectations of gifted parents that their children will inherit all their powers, it no less discountenances extravagant fears that they will inherit all their weaknesses and diseases." In other words, chances are that bright parents will have more average children and that less intelligent ones will have children brighter than they are.

Of course, as de Moivre so carefully pointed out, variations occur in human populations. His observations relate to the average across an entire population; there are plenty of individual cases that seem to defy Galton's assertions. He pointed out several that can account for variations in human populations: genetic anomalies or the phenomenon of assortative mating. The latter simply means that like people are attracted to like people. For instance, I'm six foot three and married to a woman who's tall as well (she's five foot nine). We both bring unusually tall genes to our children, so our children probably will grow to be quite tall as well. However, Galton discovered, children of tall people tend not to be as tall as their parents. His reasoning explains why the world is not filled with basketball players and midgets, with successive generations of people magnifying their parents' traits. (It is true that the American population has been gradually growing taller, but that change probably has to do with better nutrition.)

Significantly, Galton not only discovered the regression to the mean, he also recognized that it was not random or arbitrary, but systematic. In the case of the peas, he found that the difference between parent and daughter varied proportionally regardless of their size. In other words, the peas showed the same statistical variability. With help from mathematicians, Galton described the nature of this variance in a single number he called the regression coefficient.

Galton's discovery of regression led him to another groundbreaking finding, a concept that has become a foundation of modern risk analysis. As with regression, other mathematicians would work out the details. Galton's brainchild was the idea of "correlation," the idea that events or conditions can be tied to each other through connections far more subtle and indirect than straight causation. Understanding correlation permits us to peer further into the mysteries of how things happen and the nature of risk.

The most direct relationship between two events is causation. Causation is simple enough: swiftly swing your foot in contact with a soccer ball and the ball will move. Or, pour enough noxious chemicals into a small lake and the fish in it will die — simple cause and effect. Correlation describes less clearly defined relationships and associations. Take cold weather and snow, for instance. A causal relationship between the two would suggest that cold weather causes snow. Obviously, however, snow doesn't always fall when the weather becomes cold. Snowfall occurs when a front brings moisture into an environment where the conditions are right. Those conditions include cold weather. While the relationship between cold weather and snow isn't causal, some kind of relationship does exist. These kinds of relationships outside the realm of direct causation are known to scientists as correlations. Working out correlations between all sorts of seemingly unconnected events has opened up new vistas across numerous scientific disciplines. Thus, one can measure the correlation between summertime heat and the murder rate or the gold standard and the health of a nation's economy.

Working with correlations is a double-edged sword. The *Harvard Public Health Review* reports a correlation between height and the incidence of certain kinds of cancer. Tallness is a risk factor for breast, prostate, and colon cancer. But are the cancers caused by being tall itself or some other factor? Many correlations turn out to be false — they suggest an association between events where in fact there is none. If you go out and shoot pool, you'll notice that pool halls are full of beer drinkers. To a person who'd never seen the game of pool, there might appear to be a correlation between drinking beer and playing pool. One could surmise that to play pool one might need to drink beer. Obviously, this is not true. Both activities are relaxing; that's why they're often done together. Sorting out what's a true correlation and what's not is pivotal to modern risk analysis and determining where true risk lies.

In many instances, people intuitively sense the difference between causation and correlation, without needing to know the terminology. But like his predecessors, Galton managed to push beyond intuition. He established a way of measuring the extent of correlation between objects or events. This would have enormous implications. In an 1888 paper, he used the example of the length of a man's arm and his leg as an example. Things are said to be correlated, he wrote, "when the variation of the one is accompanied on the average by more or less variation of the other." General observation of the human form reveals that the length of a man's arm is correlated with the length of his leg, that is, a person with a long arm usually has a long leg as well. But, he asked, how close is that correlation? "If the co-relation be close, then a person with a very long arm would usually have a very long leg; if it be moderately close, then the length of his leg would usually be only long, not very long; and if there were no co-relation at all then the length of his leg would on the average be mediocre." Galton argued that a strong correlation — very long arm usually going with very long leg — suggests that the variations in these two features share a common cause. No correlation at all would suggest no common connection. In between the extremes of absolute correlation and none at all lies a wide range of stronger and weaker connections. Galton determined to figure out a way to quantify the strength of the correlation between two different events or objects.

One day, while taking a walk along the grounds of Naworth Castle, Galton experienced a burst of insight, the classic lightning bolt of realization commonly recorded in the history (and myth) of science. "A temporary shower," he wrote, "drove me to seek refuge in a reddish recess in the rock by the side of the pathway. There the idea flashed across me, and I forgot everything for a moment in my great delight." His insight was rooted in his earlier concept of regression, that is, establishing a single number (the regression coefficient) to describe the amount of variation in a generation of sweet peas. Galton realized that he could apply this same idea to two entirely different events and determine their degree of connection, that is, the correlation between them. In other words, he could describe mathematically the amount of variability found in each event. When two events feature a similar degree of variability, a strong correlation exists between the two. The closer the similarity, the closer the correlation. Galton's insight opened the way for science to examine the possible connections between all sorts of different things, regardless of whether their units of measurement were similar or not.

Instead of just figuring out causation, science can now explore the impact of quite subtle risks. This refinement has given birth to, among other things, the modern term "risk factor." Today, it's common practice to distinguish between health risks and risk factors. A physician might explain that a diet high in fat increases the risk for heart disease. This is simple cause and effect: fat can clog arteries, leading to arteriosclerosis, and blockage of the blood pumping through the heart. Therefore, a high-fat diet is a health risk. Obesity and high blood pressure, however, are *risk factors* for heart disease. No one can yet prove that these conditions directly cause heart problems. Research shows strong correlations — obese people and people with hypertension suffer heart attacks more frequently than people who are not overweight and have normal blood pressure. By treating one of these risk factors, individuals can lower their risk for heart disease. Determining these more subtle signals gives researchers the ability to look more deeply into health risks and determine what preventive actions can minimize the risks. For example, recently I was diagnosed with mild hypertension, a condition that runs in my family. Knowing that this puts me at risk for stroke and heart attack in the next decade or two, I started taking medication. The knowledge of this risk factor — and the preventive actions I've taken — may extend my longevity well beyond what it would have been without such knowledge.

Yet even strong correlations can also spread confusion among the public. If a study associates ingesting coffee, say, with a certain type of cancer, a journalist will often report the link, but not bother to note that the study found correlation, not causation — and a weak correlation to boot. An individual eager to determine what this information means personally will simply assume that the study shows direct causation and come to believe that avoiding coffee will prevent cancer. In fact, when one study noted only a weak correlation between coffee and pancreatic cancer, scientists conducted further studies to discover whether the beverage and the disease were actually directly linked. They found that they weren't, but the correlation still exists.

The work of Pascal, de Moivre, Graunt, Galton, and others in probability and statistics blossomed into thousands of different applications. As economist Peter Bernstein has noted, the breakthroughs in probability theory provided the structure to look at the world with fresh eyes. He writes: ". . . without that structure, we would have no systematic method for deciding whether or not to take a certain risk or for evaluating the risks we face. . . . We would have no way of estimating the probability that an event will occur — rain, the death of a man at 85, a 20% decline in the stock market, a Russian victory in the Davis Cup matches, a Democratic Congress, the failure of seatbelts, or the discovery of an oil well by a wildcatting firm." The sophisticated new tools in probability and statistics brought new shape to forecasting and introduced a powerful new way of analyzing risk. They proved critical to the insurance industry; the emergence of the stock market; and the possibility of large public works, such as railroads and skyscrapers, that no single company could build alone unless the risks were spread and outcomes carefully forecast. Farmers could stay in business even after a period of catastrophic weather. In epidemiological studies, scientists began to understand the roots of sickness and disease. In some form, probability and statistics have played a critical role in all the advances that have brought us longer lives, better standards of living, and less vulnerability to the ravages of nature. They remain among the most powerful intellectual tools that humankind has ever invented.

Paradoxically, while experts have become well versed in using these tools and explaining the results to the public, most individuals have yet to absorb these methods and use them in their own lives.

As we'll see, even the rudimentary use of probability theory and statistics in everyday life can bring about remarkable new ways of thinking about risk and making decisions. Once an individual starts looking at his or her own world using these tools, it will never look the same again.

CHAPTER THREE

EVEN LYING IN BED IS RISKY

After a few months of reading risk statistics, I had a curious experience one morning, an epiphany of sorts. At the time, however, I felt more like Alice in Wonderland after taking a sip of the "Drink Me" potion. When I opened my eyes in bed and began to contemplate my day, I began to see it not in terms of what I had to accomplish but in terms of the risks that I would encounter. The world suddenly started looking different.

First thing every morning, I usually turn on the old green metal standing lamp that once belonged to my wife's grandmother. From my reading over the past few months, I recalled that each year at least sixty Americans electrocute themselves fatally on domestic wiring and appliances. I calculated my odds over a year as 1 in 4 million. As usual, I fantasized about remaining in bed that morning and sleeping in. Another risk tidbit popped into my slowly awakening consciousness: that lying in bed is, literally, risky. The headboard could fall, or I could fall out of bed. Each year, beds, mattresses, and pillows injure more than 400,000 Americans. Over the course of a year, I stand a 1 in 650 chance of being injured this way. My risks from lying in bed don't come just from falling headboards: sociologists have compiled risk information indicating that unemployment—what I face if I stay in bed—is bad for my health, placing me at higher risk for high blood pressure, high cholesterol, heart problems, and more. Another couple of minutes in bed surely wouldn't hurt, I rationalized. My thoughts of an amorous interlude with my wife brought to mind a recent study in the *Journal of the American Medical Association*. Passing from a relaxed state quickly to the strenuous exertion of passion can shock the body, raising my chances of a myocardial infarction by 1 in a million.

Finally rolling out of bed and stumbling to the bathroom, half-asleep, I remembered reading statistics about people who slip and fall in the home: 8,500 Americans die and nearly 2 million Americans hurt themselves each year at home. I computed my personal risk of dying from a fall as about 1 in 30,000, although my actual chances are less because I'm under seventy, healthy, and sober. As I made it safely to the bathroom, I approached the toilet and sink warily. They injure more than 60,000 Americans every year (my odds: about 1 in 4,500). The shower claims another nearly 170,000 injuries every year, shaving hurts another 40,000. I stood a 1 in 7,000 chance of cutting myself badly enough with my razor this year to seek medical attention. And if I cut myself slightly, I can't afford to become angry. Some studies suggest that getting mad increases the chances of having a heart attack.

Even the act of getting dressed places me at risk. Zippers, buttons, and other articles of clothing hurt more than 140,000 Americans each year. Jewelry-related injuries send some 55,000 to the doctor or clinic. When I open my wallet and count my change, I'm reminded that half of the paper currency and coinage in our pockets carry infectious germs. According to National Safety

Council statistics for 1996, I have a 1 in 36 chance of becoming disabled for a day or more by an unintentional injury sustained at home. In fact, I stand a greater chance of suffering a disabling injury at home that at work and in motor vehicle accidents combined.

This outlook on my morning may seem paranoid. Indeed, it probably arose from reading too many risk statistics the day before. However, the fact that I can examine my morning ablutions so completely in terms of risk and probability is something new in human history. Worry, of course, is not new. "We are, perhaps, uniquely among the earth's creatures, the worry animal," wrote biologist and keen observer Lewis Thomas. "We worry away our lives. . . ." What's unusual is that I'm able to quantify—put numbers on—the concerns I have about my health and mortality. We've always lived in an uncertain world, but it's a new twist to view uncertainty and the hazards it contains in a sturdy, numerical context. A single fact, such as the risk of cutting myself shaving, means little on its own; taken altogether, however, the risk estimates, comparisons, and statistics now available force us to look at our world with new eyes. The risk information, whether couched in numerical form or in the specific lingo of probability—weather forecasts of rain squalls, a new health study reporting the correlation between indoor pollution and cancer, or the dangers of a particular stock portfolio—is the manifestation of a remarkable recent change in how experts use probability to analyze uncertainty and understand risk. If we as individuals can start to incorporate some of these ideas into our own lives, the results can be quite dramatic.

While Blaise Pascal's work on probability brought about a major revolution in thinking about uncertainty, it's a mini-revolution that began in the 1970s that brought our understanding of risk into the modern age. This revolution has to do with the emergence of the new science of risk analysis, a multidisciplinary, largely academic field of scientific endeavor. The science arose from the confluence of several factors: a critical accumulation of data in health and safety matters, the introduction of high-speed computers that could contain and process this information, and the development of sophisticated analytical techniques to work with this information. Insights from this burgeoning science have seeped into mainstream thinking largely unannounced, and now risk numbers and comparisons are ubiquitous. As we'll soon see, the fact that I can look at my morning through the prism of risk reflects a new perspective that carries with it vast implications for how I order my priorities, make decisions, and ultimately choose to live my life. It's a revolution that affects each and every one of us. In essence, we as individuals can use these new intellectual tools and perspectives to become our own best risk analysts.

In 1979, Harvard's Richard Wilson published a brilliant paper in the journal *Technology Review*, a landmark in the science of risk analysis. "To compare risks," he wrote, "we must calculate them." Wilson went on to calculate a series of everyday risks that increase a person's chance of death by 1 in a million. Two dozen different activities are on his list; all the activities raise an average person's chance of dying by 1 in a million: smoking 1.4 cigarettes (cancer and heart disease) or spending 1 hour in a coal mine (black lung disease). Others on the list are less intuitive: traveling 6 minutes by canoe (accident), eating 40 tablespoons of peanut butter (liver cancer from aflatoxin B), living 2 months in the average stone or brick building (cancer caused by natural radioactivity), and living 2 months with a cigarette smoker (cancer, heart disease).

As Wilson points out, taking these figures too literally is dangerous. A person doesn't decide to eat only 39 tablespoons of peanut butter, for instance, to avoid the risk of aflatoxin. These numbers represent averages only. However, Wilson's work reveals two important insights. First, disparate activities can be compared using a numerical assessment of risk as the common denominator. Second, a scientist can define an absolute risk threshold — in this case, 1 in a million — in terms of everyday activities. We all have at least some experience with Wilson's everyday examples. From these common experiences, we can form an understanding about what constitutes such a risk. It's then possible to work up to more serious risk ratios and understand their relative likelihoods.

Decades after Wilson's work, the average person finds himself or herself barraged with risk information of all sorts: what's healthy and what's not; how the environment is deteriorating; how long we can expect to live; whether it will rain or we will die in a car wreck. We juggle this information every day, trying to make sense of it and figure out how it applies to us. Part of the reason for our confusion today over what's safe and healthy, as reports contradict one another in the press and information grows exponentially every day, is that we're relatively new to dealing with risk information of this complexity and volume. At the same time, the media, scientists, bureaucrats, regulators—the experts—are unused to presenting risk information to the public.

I think of the phase we're in now as the wild west period, because risk information and stats fly around the media like errant bullets, often ill defined and vague. Most often the information leaves us confused and in need of more information to put the data into some kind of relevant context. But sometimes there are heady moments when we come across a piece of risk information that gives us an inside track on how to negotiate the hazards of life. When I was considering staying in bed, I remembered University of Pittsburgh physicist Bernard Cohen's calculation that the risk of being unemployed is roughly equivalent to the risk of smoking 10 packs of cigarettes a day. Other pieces of information prove equally intoxicating: *Discover* magazine reported that being born male and proceeding to do typically male things, an individual loses 2,700 days of life. Smoking a single cigarette, various sources reported, knocks as many as 12 minutes off a person's life. Six seconds disappear off a person's life for every driving trip taken without a seat belt. Consuming a diet soft drink knocks off another 9 seconds.

Massachusetts Institute of Technology's Arnie Barnett calculated the risk of flying in a commercial airline by suggesting that an individual would have to take a flight every day for the next 26,000 years to experience a crash. Few of us haven't come across a piece of information recently that sticks in our craw, causing us to reevaluate the wisdom or safety of something we do regularly. For me these facts are electric because each seems to offer a tiny window onto my mortality, as though I can peer into my future.

What makes risk analysis so topical and exciting today is that science and society have accumulated enough data in safety, health, and other fields to make any number of risk comparisons possible. Interpretations of information, of course, have been rampant in human history; but only recently have enough data been compiled to give interpretations the full weight that the science of probability and statistics can impart to them. Consider, for example, the realm of traffic safety. Since January 1, 1975, the National Highway Safety Administration has logged details about every motor vehicle fatality into an immense computerized database, known as the Fatality Analysis Reporting System. It contains the records of more than one million deaths, a grim reminder of the deadliness of America's favorite mode of transportation. In an average year, between 40,000 and 45,000 people die on the nation's roads. The data logged about fatal accidents use information that ranges from the speed and make of the vehicle to the time of day and age and sex of the driver. From this database, traffic safety scientists have characterized the risks of driving. Leonard Evans of the General Motors Research Labs has calculated that in a head-on crash between a heavy and a light car, the driver of the smaller car will fare worse, a matter of common sense. But, Evans's risk analysis reveals the exact numeric risk between the two: the occupant of the lighter car is 17 times more likely to die. "Choosing a car half as heavy," writes Evans, "doubles your fatality risk per year if you continue to drive the same number of miles in the same way." He notes drily, "The effect of mass is large."

Evans also calculated the relative safety effectiveness of seat belts versus air bags. Seat belts reduce driver fatality risk by 42 percent; air bags, when used with a belt, reduce it another 9 percent. This type of risk analysis enables social scientists to fine-tune the nature of risk-management strategies. How effective are seat belts? The U.S. Department of Transportation's National Highway Traffic Safety Administration (NHTSA) estimates that the lives of 9,601 people would have been saved in 1997 alone if all passenger vehicle occupants over age four had buckled up. If all motorcyclists wore helmets, NHTSA calculates, 266 lives could have been saved.

Another safety expert at General Motors, Richard Schwing, calculated that 34 percent of all fatal accidents occur in the hours between 11 PM and 5 AM, during which only 5 percent of travel takes place. Again, intuition and common sense suggest that driving late is inherently dangerous—due to a rise in the number of inebriated and/or tired drivers, for instance—but the numbers give us hard information. Avoid driving after midnight on Friday and Saturday, and you reduce your risk of a fatal accident by 20 percent.

The same goes for health issues. Long before large epidemiological studies convincingly demonstrated the harm of smoking, people understood that cigarettes were hazardous. The World War II sobriquet for cigarettes, "coffin nails," adequately testifies to this. Yet only the accumulation of statistics revealing that smoking increases a person's chances of developing lung cancer by as much as 3,000 percent revealed the magnitude and severity of smoking as a health risk. This risk information helped to elevate smoking to the category of a grave public health concern, and it proved a major factor in bringing about both remarkable changes in individual behavior and national and local antismoking legislation in recent years.

Large data sets chronicle nearly all aspects of harm that can befall an American. The National Safety Council publishes an annual booklet, *Accident Facts*, that lists hundreds of ways in which Americans die accidentally. In light of the trepidation with which I now greet my morning, I can consult this book and learn just what is the most dangerous aspect of life at home. Without access to this risk information, I would put fire, or perhaps the poisons under the sink, or maybe the microwave or garden tools, at the top of my list of household dangers. However, *Accident Facts* shows that all of these combined don't approach the risk of injuries caused by falls on stairs or from landings.

The fruits of risk analysis, bolstered by years of data, often force us to reexamine aspects of our lives that we didn't think much about before. The specific risk information, whether about the dangers of a lightweight car or stairway falls, encourages us to abandon our usual techniques for thinking about risk—intuition and common sense, for instance—and deal with new knowledge contained in the often unfamilial form of numbers, risk ratios, and comparisons. This hard-edged information clarifies the risks we face in new ways. Bringing the tools of risk analysis into our lives is not just a matter of reading and accumulating risk statistics. While often fun to read about, risk facts are like candy; they taste sweet but fade from memory quickly. Where risk analysis gives us sustenance is in helping us recognize that the information per se is not an endpoint, but pieces of a much larger puzzle. If we can be critical of this information, yet learn to trust what it can tell us and use it as a tool, it can give us great power in our lives.

A baseball fan provides a fitting analogy for how risk analysis prompts us to shift our viewpoint. A person interested in his or her home team has access to mountains of information about its success in hitting and pitching, not only during the current season but for years past. Individual batting averages, earned run averages, and team fielding statistics give a good indication of the fortunes of a team. Obviously, a batting average, be it good or bad, will have little bearing on a fan's ability to guess how a batter will hit on his next at bat. As GM's Richard Schwing has written: "There is no way to deduce a particular person's chance of fatality in the next mile he drives. Rather, one describes the exposure and then uses the observed risk for all identically described exposure as an estimate . . . it gives a perspective . . . of the risk." Past performance is not always a good predictor of future success or failure. Yet overall, a player's stats give us the ability to compare him not only with his team members and others in the league but also with those who played decades before him. A player's stats give fans a benchmark, a clear idea of how he ranks. A person takes notice if those numbers are stronger than most, or if they indicate that a player's contribution could be on the verge of exceptional, compared with all the rest of baseball players. No one mistakes the statistics for the game itself; they are only a way of gaining insight, understanding the complexities of the game, and making predictions and judgments about the outcomes of a season, a batter, or pitcher.

Now imagine that life itself is a game of baseball, except that mortality, health, and quality of life are the markers of success and failure. In this analogy, risk analysis provides the stats, forming the basis of the benchmarks upon which to make decisions and adjust behavior. All the risk estimates and comparisons that we read and hear about form a network of information we use to view our lives and inform the decisions we make. As students of the risk information around us, we follow the game, begin to keep track, and continually reexamine our lives in view of this new information. It could be about how our diet, our genes, or our behavior affects our health and life. Of course, the game of baseball is uniform in many ways: a clear set of rules, uniform equipment, and clear goals. The way our lives unfold, on the other hand, is a far messier affair. As biological creatures, we are constrained by many natural rules that bind all humans together. We all require food, our bodies respond to viruses, stress, and chemicals in many of the same ways; and we all die. Life's complexities obviously make risk information complex, too — in fact, it's the most complex game in town. Matters are further complicated because we are only now learning to integrate numerical risk comparisons and other risk information into our decision processes. The human mind did not evolve to work out the meaning of statistical calculations derived from techniques invented within the past several centuries. In many cases, we are loath to abandon intuition or common sense in favor of a statistic or other fact. Furthermore, risk numbers and comparisons aren't uniform, and in many cases they can be manipulated and misinform. Part of learning the language of risk in today's society is acquiring the ability to ferret out suspect information and then to make sense of the reliably vetted, accurate risk information that remains. The public can do little to guarantee that a scientist has done his or her analysis correctly, but they should monitor three critical factors in how that information is framed: the difference between absolute numbers versus ratios; the necessity of qualifying the time frame in which a risk statistic is personalized to your life; and the overall population to which the risk information refers. For any individual, not taking these factors into consideration can make risk information useless or, in worst-case scenarios, dangerous.

First, it's critical to distinguish between what absolute numbers and ratios tell us. For example, when the National Safety Council published a list of occupant fatality totals for 1994 by different types of vehicles, the list revealed that passenger cars killed the most people (24,300), followed by light trucks (7,600), and motorcycles (2,500). But although passenger cars clearly cause the great majority of the carnage on America's highways, these absolute numbers don't show the *relative* safety of each different form of transportation. When these numbers are adjusted into ratios according to the total number of vehicles registered, the picture looks quite different. In fact, the list is reversed. Passenger cars represent nearly three-quarters (75 percent) of registered motor vehicles, yet account for only two thirds (66 percent) of fatalities. Twenty percent of all registered vehicles are light trucks, whose occupants account for 21 percent of overall deaths. Motorcycles are the most over-represented in the fatality column when compared to total number of vehicles. While only representing 2 percent of all vehicles, motorcycle drivers and passengers account for 7 percent of all fatalities. When it comes to relative risk, only fatalities expressed in ratios have any meaning.

A second example of how numerical risk information needs to be used as a tool involves the issue of time frame and exposure, critical pieces of information that are often omitted when risks are presented in the media. When the risk of contracting HIV is not qualified, for instance, citing the number of cases alone is useless for the purpose of determining risk. Does the number refer to the risks incurred during one random sexual encounter? Or does it represent a person's odds of contracting HIV over a lifetime, or in a year of sexual activity? How risks are framed can make a large difference to whether a particular risk is assessed as dangerous or not. For an individual American, the average odds of dying in a motor vehicle crash every time one climbs into a car is 1 in 4 million. As Wilson's 1 in a million numbers indicate, this is a very small risk: a person is twice as likely to die from being struck by lightning over a lifetime. Yet consider that each of us takes an average of 50,000 motor vehicle trips over a lifetime. Our chances of dying in a car accident *over a lifetime* of driving are therefore 1 in 140. Looking at the long view — over the course of a lifetime — makes the risk of driving appear far greater.

A third critical factor in how the information is framed is what population the risk information uses — for example, the world population, Americans, men, women, a particular age group, and so on. The type of population addressed can have a significant bearing on how applicable the risk information is to a particular person. At the outset of this chapter, I mentioned that my chance of falling and dying in my home was about 1 in 30,000 over the course of the year.

That number includes all Americans regardless of age. However, risk stats adjusted for my age group (25–44) lower the likelihood to 1 in 250,000. For Americans between the ages of 65 and 75, however, that risk goes up to about 1 in 20,000. In many cases, the addition of criteria such as income level, race, and gender can reveal wide fluctuations from the average. We need to be savvy about what the average tells us specifically as individuals.

Risk analysis gives us the ability to look at virtually every aspect of our lives through the lens of risk. An august body of experts, recently assembled by the National Academy of Sciences to study risk, developed a simple, seemingly self-evident statement that became the linchpin of their multivolume report: "There is no activity, process, or product that is free of risk." All the work catalyzed by Pascal and insights later garnered in the science of risk analysis lead to this deceptively simple statement. In this lies the key to how our worldview has changed. Before I started research on risk, I never worried much about risks outside the adventures and expeditions I organized. Like most Americans, I generally perceived risk in terms of dangerous activities involving a chance of spectacular death or financial speculations such as sinking capital into a new business venture. It makes sense that with life in balance, risk has hitherto been associated with dread and those large, dramatic events that can suddenly extinguish life or cause financial ruin.

The word *risk*, however, now denotes a far more subtle range of loss possibilities. Risk analysis introduces the awareness that every time you climb into a car, take a drink, light the barbecue, go running, have sex, or make any of the thousands of other decisions you make every day, you take a risk. But it offers far more than just this simple recognition. Risks can flow from conscious actions (shaving, driving, hang gliding) or can be incurred in situations where people do *not* have a choice (breathing polluted air, exposure to radon, the expression of a genetic disease, or getting crushed by a falling aircraft). Risk has a time dimension: the loss or harm of a particular risk can be immediate (a car wreck) or chronic (the emergence of a fatal melanoma from decades of exposure to the sun). An environmental risk, such as global warming, may take centuries before it begins to cause detectable harm to the planet.

Risks exist both when an individual performs an action and when he doesn't. A sick man can choose to heed his doctor's advice to undergo a minor operation. He incurs risk if he does so (the drive to the office, the possibility of the procedure going wrong) or if he doesn't (the condition could get worse, become not treatable, and threaten his life). The definition of what constitutes risk has exploded so quickly and furiously that the word itself can no longer comfortably express its myriad meanings and connotations.

Every decision between at least two alternative courses of action holds out different risk scenarios. Thousands of such decisions fill our everyday life. Crossing the street, for instance, when you are late for an important appointment. Do you cross the street now with oncoming cars in the distance, or do you wait? Near instantaneous cost-benefit calculations flash through your head. A dash across the street can save you several minutes but will also increase your risk of getting hit by a car. Conditions complicate the decision every time: is it raining, how important is the appointment, is the street crowded with rush-hour drivers racing to get home? Consider an interesting observation to come out of risk analysis. More people die while crossing the street at crosswalks than while jaywalking. Of course, we should exercise care with this piece of risk information, because it reflects absolute numbers and not a risk ratio. More people cross the street at crosswalks than not, so the opportunities for crosswalk accidents are greater. However, this statistic could indicate that particular dangers exist for pedestrians at crosswalks because they don't pay as much attention to traffic patterns as a person crossing against the traffic.

The vast majority of day-to-day risk decisions don't merit conscious evaluation. Although you may realize that there's a small risk of an accident driving to the grocery store, it certainly doesn't overshadow the need to feed your family. Nobody even thinks twice about a decision like this. Of course, many of the risks of life are inconsequential. None of the statistics about electrocution by bedside lamps or slipping on the bathroom floor would prevent me from getting out of bed. After all, I've routinely faced far worse physical risks while ice climbing or running rapids. Of course, winging it, guessing, and using intuition and common sense get us through most of life's myriad decisions just fine. Stephen Breyer, now a Supreme Court justice, once wrote that many decisions in life are like canoeing down a river. The canoeist doesn't need to know the exact contours of the riverbed to navigate the river. Often, he simply heads into the strongest rush of water, lines up through the center of the channel, and gets through safely. An individual can finesse many risk decisions.

Problems arise, however, when the river drops into white water, forks, or presents several possible routes to go down. Which route to take? Is it time to portage and line the canoe or run the rapid and set up safety ropes below it? The complexity of modern life offers many more rapids in the river. When confronted with a recently released medical study, should one undergo a certain operation, or change behavior? Risk information gives us tools but forces difficult decisions as well. "Faced with such complex problems," writes risk expert Baruch Fischhoff of Carnegie Mellon University, "we usually just muddle through. We have in place a set of practices that have evolved over time. Some, we have adopted deliberately (e.g., low-salt diets). Others were imposed upon us (e.g., automatic seat belts). Still others were copied from friends with little attention to safety (e.g., smoking) or are of uncertain origin (e.g., triple-checking the stove before leaving the house)."

In certain cases, listening to the new findings in risk analysis reinforces our common sense and gut feelings. In other cases, however, they show that our intuition is just plain wrong. I have a friend who is deathly afraid of flying and drinks heavily to numb his concerns before he boards. As most people know by now, my friend essentially has nothing to worry about when he flies. Ninety-three Americans die on average each year in commercial airline crashes. Even if the fact that only some Americans fly is taken into consideration, the major causes of fatality for all Americans at home, work, and play in 1996—falls (14,100), drowning (3,900), poisoning (10,400), and vehicle accidents (43,300)—remain far more significant (and many involve the abuse of alcohol). Yet it's not productive to criticize my friend's fears, even if they are irrational. However, I grit my teeth when I see him riding his bike all around my busy suburb without a helmet. According to the U.S. Department of Transportation's NHTSA, 813 bicyclists died in 1997. Seventy-five percent of those deaths, reports the National Safety Council, result from head injuries. Compared to flying or even driving, bicycling is a far riskier undertaking, even allowing for the health benefits of the exercise. In fact, bicycles are the most dangerous form of transportation per mile traveled. Philosophically, I can think of my friend and others who don't ride with a helmet as either ignorant about how dangerous this practice is or stupid because they know the risks but choose to ignore them. I do think that individuals act irresponsibly and stupidly quite often when it comes to making good risk decisions. However, overall, I firmly believe that they do so more out of ignorance than out of stupidity. I see that learning the language of risk is a critical way to elevate ourselves out of ignorance to make good and informed decisions in our lives.

Looking at the world through the lens of risk brings into sharp relief the fact that the very act of defining risk and striving to quantify it forces us to consider the trade-offs inherent in our decisions and to realize that these decisions involve clear statements about our values and beliefs. Suddenly, we must answer certain kinds of questions we might have ducked before. Traditionally, mortality statistics have been listed as absolute numbers and ranked accordingly.

By this method the leading cause of death today is heart disease, followed by cancer; accidents trail a distant third. These figures do not differentiate between the death of an eighty-five-year-old man and that of a ten-year-old boy. They are simply tallies. However, the old man lost zero potential years of life, because he had already outlived the average life expectancy. At only ten, the boy died more than sixty years before his average life expectancy.

To attempt to make mortality statistics more meaningful, risk analysts now factor in loss of average life expectancy as a tool to examine mortality information. Using this technique to evaluate the top three causes of mortality in the United States, the list is turned on its head: accidents significantly outstrip cancer and heart disease in taking away potential years of life. As the leading cause of death for young people, accidents cause the greatest potential loss. These issues bring difficult questions into focus. While spending money on heart disease prevention may save more lives, it may only extend these lives a few more years. Accident prevention would save fewer lives overall but would extend the lives of those it saved by many more years. Yet the loss of average life expectancy in turn minimizes the impact of diseases, such as cancer, that take a long time to develop. The very process of defining and trying to quantify risk reveals subtle and not so subtle valuations and judgments. Is the life of a young boy more sacred than that of an old man? How much is a life worth? Or an ecosystem, a species?

As we embrace risk analysis in our individual lives, we face many personal decisions about what do with the risk information we encounter. Since falls are the leading cause of mortality in the home, should we carpet our stairs or forgo positioning a picture on the wall at the bottom of the stairs? Should we be more careful after we drink alcohol, install better lights, caution the children against roughhousing on the stairs, recommend single-story housing for elderly parents? When we buy a new or used car, how do we use the information that a person's risk of dying in a car accident doubles per year if he or she drives a light car versus a heavy one? Looking at our lives through the lens of risk gives us an entirely different vantage. It enables us, for instance, to understand how risk accrues over the course of our lives, not just when we happen to step into a car or board a plane.

Armed with information such as this, we are ready to become much better risk analysts in our own lives.

CHAPTER FOUR

RISK MANAGEMENT'S ORWELLIAN SIDE

It had been a long night caring for my sick six-year-old son, laid low by an earache. The next morning, bleary-eyed, I rang the doctor. He gave me a prescription for the recurring condition. I headed out to the pharmacy. Upon climbing into my Chevrolet Cavalier station wagon, a buzzer assailed me because I was slow to put on my seat belt. On the road, speed bumps jarred my tired body, forcing me to drive more slowly as I navigated the side streets to the pharmacy. The happy thought of eating a jelly doughnut at the bakery next to the drugstore evaporated when a public service announcement came on the radio and lectured me about the dangers of fat. Back at home in my somnolent state, I couldn't seem to open the childproof top to the antibiotic. After a few minutes of struggle, the top came off, splashing a bit of the sticky pink medicine onto my pants. Moments later I engaged in a tug of war with the tamperproof wrapper on the Tylenol.

During my morning adventure, I had had the distinct impression that a clumsy guardian angel was watching over me, interfering with every move I tried to make. With my son now sleeping peacefully, and after I had drunk another cup of coffee, I realized that despite my tiredness, the gauntlet I felt I'd run was worth the inconvenience and hassle. Indeed I *should* wear a seat belt in the car; I *should* watch my fat intake. The childproof top that had nearly foiled me *would* prevent my curious children from ingesting an overdose of medicine, and for that I'm glad. And I felt safer knowing that the Tylenol could not have been tampered with by some consumer terrorist between the factory and the drugstore. Still, these realizations didn't assuage my irritation at being second-guessed by mechanical contraptions, warnings, and the like.

As I realized that morning, we live, work, and play in a world constrained by a tangled web of voluntary and involuntary safeguards, all designed to minimize physical risks. Over the past century, risk practitioners from highway engineers to consumer product designers and government bureaucrats have used probability theory, data-crunching techniques, and sophisticated information gathering methods to assess the consequences of hundreds of thousands of everyday activities. As a result, piecemeal strategies have been legislated, mandated, and implemented to moderate these hazards. In modern risk analysis parlance, these attempts to master and eliminate risk come under the heading of "risk management." Thus in the case of children's medicine, plastic bottles replace glass, easy-to-open lids surrender to childproof caps, and child-appropriate doses are carefully established. Children's pajamas become primarily polyester as federal laws regulate the sale of more comfortable but far more flammable cotton products. The practice does work. In 1970, two years before institution of these children's sleepwear regulations, sixty children died after their nightclothes caught fire. In 1991, only six children died in such incidents.

Wherever we look, these risk-management strategies are

present: sidewalks prevent pedestrians from being hit by cars; individual sugar packets at restaurants stop contamination from large, open containers; smoke detectors warn families of fire; passive restraints and air bags save thousands of lives; nutritional labels on food products guide consumer eating patterns toward greater health; sophisticated tornado and hurricane alert systems on the radio warn vulnerable populations in time to evacuate.

Within the past fifty years, in millions of instances, society has traded money, convenience, and some degree of freedom for greater safety in people's lives. Added together, these risk-management techniques and devices seem positively Orwellian—a benign but all-powerful presence in our lives, subtly guiding and shifting the way we act. For the most part, a safety cap or speed bump hardly registers in the hustle of modern life. However, all told, these devices have profoundly altered the way we interact with our environment and with risk. So powerful has the effect of risk management become that it has qualitatively changed not only our behavior and outlook on risk but also the very nature of the risks we face. The vast bulk of these changes have occurred in the last quarter century at a speed so dizzying that they have led to confusion and greater worry, despite the fact that our lives are growing safer. Americans, including most experts, are still struggling to understand what risk management has wrought.

Of course, the effort to reduce risk is nothing new in human history. In the same way that individuals didn't need the latest insights in risk analysis to identify hazards, they didn't wait until the development of risk management to try to reduce them. Many of the major advances in civilization can be viewed as risk reduction strategies on one level or another. Hunter gatherers, for instance, banded together in part to maximize their access to food. A hunter knew that while he might have to share the results of a hunt with others, he would also be assured of food should his individual hunting efforts fail.

Banding together also provided protection for those who grew old, infirm, or weak. The development of agriculture and domestication of animals provided stable food sources during the winter or in times of famine and want. Until all the innovations cascading from Pascal's discovery of probability, however, humankind's dealings with risk were limited to the reactive and personal. People went about their lives trying to keep risk to a minimum, but a certain amount of what happened to them was inexplicable and therefore, it seemed, not under their control. When the remarkably powerful new tools to quantify risk were unleashed, it became possible actively to manage risks, not just to respond to them. This process has culminated in the late twentieth century in the systematic, national management of risk.

Pascal's theory of probability added to a world exploding with new information about the way the world worked. During the Renaissance, revolutions in science, religion, and political theory freed the individual to think of himself as a free-acting entity, capable of exercising great control over his future. At the same time, the scientific revolution of the early Renaissance established the underlying philosophy and process for determining how and why things worked. From this great rush of discovery emerged the idea that scientists could find out the reasons for everything, simply by looking long and hard enough. Science began to reveal that events and conditions once attributed to chance, the gods, or magic, were in fact caused by tangible natural forces and substances. Probability found applications across the sweep of life in evolutionary biology and at the atomic level in quantum physics, theories that strive to describe a web of causality in terms of uniform principles. Probability and chance, it now appeared, were woven into the fabric of the universe. Freed from the yokes of a deterministic worldview, humans could take bold steps to vanquish hazards that had long plagued them.

Just one of many advances in medicine reveals the power of risk management over the fabric of society. It started with a remarkable Englishman of wide-ranging interests who lived in the late eighteenth and early nineteenth centuries. Edward Jenner's eclectic interests included the biological world — he wrote a book on the natural history of cuckoos — but his fame rests soundly on the work he did that led to the eradication of one of the worst diseases of all time: smallpox.

While bubonic plague periodically swept across Europe and killed people in droves, smallpox was a steadier and more constant killer across the ages. Anyone who lived into adulthood had already experienced several smallpox epidemics. Once exposed, some people contracted it and died; others lived and became immune for the rest of their lives. Those who died first developed a fever; next the fateful pustules erupted on the victim's face; then the patient vomited blood as a result of internal bleeding. The disease killed and scarred millions of people all over the world. Smallpox came to the New World in the sixteenth century in the body of an African slave aboard the ship of a Spanish conquistador. It spread like wildfire, killing untold numbers of Aztecs, Mayas, and Incas. The weakened Meso-American cultures could do little to resist the new disease. During our own century, according to Michael Oldstone, a virologist at Scripps Research Institute, smallpox may have killed as many as 300 million people.

For centuries, healers had struggled to quell the bane of smallpox. Chinese healers in the tenth century took dried scabs from smallpox victims and blew them up the noses of unexposed people. Unfortunately, this is one of the paths through which smallpox naturally spreads, so the potential of infecting those they were trying to help proved high. This approach just speeded up nature's work, killing those not immune and strengthening those who were. Early in the eighteenth century, a Boston physician named Zabdiel Boylston inoculated patients with the pus from smallpox lesions of afflicted people. He believed that an inoculated person would develop a mild case of smallpox but soon recover and so develop a lifetime immunity to the virus. His work provoked a furor. The idea of injecting a person with smallpox itself flew directly in the face of medical convention and oath. His patients survived better than those he didn't treat, but some still died from the process itself.

Jenner's extraordinary breakthrough, one of the greatest advances in medicine to date, came seventy-five years later. In 1796, Jenner began conducting experiments based on an intriguing story he had heard from milkmaids living in Gloucestershire. He had learned that milkmaids who contracted the virus cowpox from milking infected cows did not seem to get smallpox. Because cowpox is harmless to humans, it gave Jenner an idea. From cowpox serum, Jenner created a smallpox vaccine that he first gave to an eight-year-old boy. Weeks later, he inoculated the boy with smallpox. The boy didn't contract the dreaded disease. To many people across the world, Jenner's discovery would seem more miracle than science. However, Jenner's success lay firmly in the application of the scientific method. It was a magnificent time for science and marked a turning point in humanity's ongoing struggle with nature. Instead of just minimizing risk, Jenner originated a tool — inoculation — that others could use to manage the spread of smallpox. Inoculation would prove one of the most effective risk-management tools of all time, the key to fighting not only smallpox but also measles, mumps, tetanus, chickenpox, cholera, diphtheria, bubonic plague, influenza, typhoid, yellow fever, and others.

The story of smallpox ends not quite two hundred years later with one of history's greatest public health victories. In 1966, a medical team from the United Nation's World Health Organization (WHO) undertook a ten-year international effort across the world to eradicate the disease. Hundreds of millions of people received inoculations with new mass vaccine techniques. In this intensive effort, the WHO teams finally tracked down the last naturally occurring case of smallpox in the world. The patient's name was Ali Maow Maalin from the Somalian town of Merka, and he survived. By

1979, after a small outbreak from a British laboratory was quelled, smallpox had been taken out of circulation completely. Today, smallpox sits quarantined in the freezers of scientists, locked away in vials like an evil genie stuck in a lamp. In less than two centuries, science eliminated one of the worst scourges the world had ever known. Paradoxically, the only risk from smallpox now comes from terrorists and state-sponsored bioweaponeering programs that aim to combine smallpox with other viral killers.

Within the time of Jenner's discovery and the eradication of smallpox, science also developed germ theory, the underlying basis for how disease is transmitted at a microscopic level. From the work of the French chemist Louis Pasteur and others emerged scores of risk-management strategies. Scientists learned that heating milk killed most microorganisms; that clean drinking water stopped the spread of typhoid and cholera; that antiseptic conditions in surgery prevented infections; and that sulfa drugs and antibiotics crippled age-old killers, such as pneumonia.

Within the last two hundred years, these innovations have exerted a profound impact on the lives of individuals in the Western world. In the seventeenth century and before, getting a splinter or breaking a bone often meant the onset of infection, which commonly proved fatal. In World War II, for the first time in human history, infectious diseases killed fewer soldiers than combat did, thanks in large part to new drugs that fought infection. The WHO is now working to eradicate a second major human disease: polio. Eighty-two million children in India received oral polio vaccinations on a single day in January 1996. Measles, which still kills more than a million children a year, is targeted for eradication sometime early next century. Today, former childhood killers—whooping cough, scarlet fever, and diphtheria—are no longer major threats in the Western world.

A comparison of the mortality tables at the beginning of the twentieth century and today reveals the extraordinary impact of the advances engendered by these medical and health practices on the risks facing the average American. In 1900, the three top killers of Americans were all infectious diseases: pneumonia, tuberculosis, and "diarrhea, enteritis, and ulceration of the intestines." The only infectious disease on the list today, pneumonia, is number six. By wiping out infectious diseases, medical science made an astounding contribution to human longevity. The life expectancy of my grandparents, born around 1900, was well under fifty years. In comparison, male infants born today can be expected to live more than seventy-two years, while female infants on average will live to about age seventy-nine. (Some of these changes can be attributed to great strides in fighting infant mortality, but not all.) Another comparison puts these numbers in a different perspective: at the beginning of the twentieth century, 75 percent of the U.S. population died before age sixty-five. Today, this figure is nearly reversed: almost 70 percent live beyond sixty-five. *American Demographics* reports that a baby girl born in 1900 could be expected to live to age forty-nine; but a girl born in the year 2000 will probably live almost eighty years. My children can expect, on average, to live half as long again as their great-grandparents.

Life expectancy has increased because risk-management techniques enable people to live out their natural lives. For millennia, the natural limits of human life have remained generally the same as noted in the Bible, three score and ten years. The human body seems to hit an absolute age barrier — around 115 years. No one can prove definitively that any person has ever lived significantly longer. (It's impossible to verify the claims of those long-lived people of remote Georgian provinces of the former Soviet Union.) Humans live longer than they used to not because natural life spans have been extended, but because risk-management techniques have eradicated or weakened many threats to life.

Risk-management strategies did more than add to human longevity. They qualitatively changed the nature of the risks we face.

Current mortality tables give clues to this remarkable transformation. When inoculations, antibiotics, and other recent medical inventions effectively knocked out infectious diseases from the top spots in the mortality tables, a new order of risks took their place: heart disease, cancer, cirrhosis of the liver, and stroke. As opposed to infections that come from nature and often kill quickly, these new killers are chronic and derive largely not from nature but from habits, lifestyles, and behaviors of modern life. When these chronic killers bumped infectious diseases off the chart sometime during the 1930s or 1940s, a little-recognized but major milestone in Western history occurred. For the first time, infections no longer killed more people than war and famine.

Although the mechanisms of chronic disease, such as cancer and heart disease, are not fully understood, exhaustive epidemiological studies reveal that lifestyle and diet greatly influence their incidence and severity. J. M. McGinnis and W. H. Foege reported in a 1993 study published in the *Journal of the American Medical Association* that roughly 85 percent of known diseases causing premature death can be attributed to lifestyle choices. Cigarettes, alcohol, diets high in saturated fat, obesity, and sedentary living vastly increase an individual's chances of contracting such diseases. When death comes today, it's no longer mostly from infectious diseases but from technology in the form of accidents or other ills. Car crashes take the lives of nearly 50,000 Americans every year, gun-related incidents, including suicides, claim another 30,000 lives. Food-processing, preservation, and packaging techniques make it possible for Americans to eat high concentrations of fat, a major contributor to heart disease. The mass production and distribution of cigarettes contribute substantially to the high incidence of lung cancer that claims as many as half a million American lives a year. These are the ills of modern society, and they are of our own making. For all of human history up to now, the major risks facing humans were like a large armada anchored off our shores. Within a matter of decades, we have come to realize that the major threat no longer comes from offshore but lies inland. And our guns, long trained on the sea, have proved useless against these new attacks. In fact, American scientists are still in the process of wheeling the guns around and finding the new targets.

As risk-management strategies banish old orders of risk, the emergence of new technologies creates new types of hazards. Out of World War II came mega-industries, new industrial processes, an astonishing number of new products — and large quantities of hazardous materials. Some industries churned out synthetic chemicals and their by-products, while others gathered natural elements into concentrations never seen in nature. Transportation networks, civil engineering projects, mining, and the distribution of energy all grew exponentially.

The idea that we ourselves created our own worst hazards began to sink in. The post–World War II era began to reveal the darker side of technology. Of course, it would be misleading to suggest that no one had ever recognized the disadvantages of technology until then. The Greeks and Romans, for example, knew all too well that the lead they depended on to protect ships and build buildings, the lead they used in plumbing, weapons, and jewelry, was also the substance that sickened miners, poisoned the water supply, and contaminated wine. During the Industrial Revolution, Charles Dickens and Emile Zola described in novels the death, disfigurement, and social disruption to masses of workers caused by the factory system. Certainly, the carnage of World War I brought a chorus of anguished cries about the power of modern technology to kill on a scale never before imagined. Still, historians of technology argue that until recently these voices remained a minority, overshadowed by a majority who viewed technology with a mix of wonder and hope. The promise of a bright future with technology at the helm kept many of the critics of unchecked technological and industrial innovation at bay.

Concern about the dangers of unbridled technology developed in the United States after World War II, eventually giving birth to the consumer, environmental, and regulatory movements. Crusaders such as Rachel Carson (*Silent Spring*) and Ralph Nader (*Unsafe at Any Speed*) brought the stunning downside of scientific and technological innovation to national attention. Setting the stage for these voices were earlier government measures to protect the public. In 1945 the FDA banned the food dye Butter Yellow; in 1950 the artificial sweeteners Dulcin and P-4000 were banned, as was the use of aminotriazole on cranberries. Between 1957 and 1978, the U.S. Congress passed more than 178 laws concerning technological hazards and their management. This burst of activity established the Environmental Protection Agency (EPA), the Occupational Safety and Health Administration (OSHA), and the Consumer Product Safety Commission. The idea that technological benefits and quality of life were sometimes offset by sickness, environmental degradation, anxiety, economic costs, premature loss of life, and injury gained wide acceptance. Technology was no longer worshipped as the savior of human society. Instead, a more thoughtful and sophisticated national dialogue emerged about trade-offs and risks versus benefits. Increasingly, the word *risk* became invoked as a useful term for framing these difficult issues.

The discovery of the shape of DNA and subsequent recombinant technologies is only one example of how new technologies can bring great promise and peril simultaneously. While such benefits as artificially manufactured insulin and the defeat of genetically induced diseases became feasible, DNA manipulation also opened up the frightening possibilities that mutant strains of usually harmless bacteria would run amok or that scientists would misuse their knowledge to clone humans. Other issues were raised too: if an individual's DNA revealed a flaw, such as a propensity to develop a terrible disease, what would we do with this information, and who should know about it? Other technological breakthroughs brought new benefits and unanticipated problems. The Green Revolution of the 1960s introduced artificial fertilizers to the rest of the world, accounting for staggering increases in crop output. But now the natural cycle of nitrogen in the world is becoming overloaded and may cause major atmospheric imbalances. Pesticides promised farmers an upper hand in the fight against crop pests, but as agricultural insecticides such as DDT leached into the environment, they began to wreak havoc: massive numbers of waterbirds began to die off, while pests simply became immune to conventional pesticides. Antibiotic-resistant strains of bacteria are emerging because of the everyday overuse of antibacterial products in home, hospital, and industry. Technology and science, it seems, have moved too fast, discovering and innovating potentially dangerous materials and processes before society can wrestle down all the ethical, social, and material implications — and there's often a long lag time before risk-management technicians can address and minimize the problems. Thus, risk managers are in a constant catch-up game with technological innovations.

In the 1970s and 1980s, the new science of risk analysis emerged, making use of information and analytical techniques from across academic disciplines in an attempt to quantify the trade-offs and balances of new and existing technology. Engineers and technological experts struggled to project the possible downside of complex and brand new technologies. The American public joined the debate too and let their concerns be known. The great promise of fission nuclear power as a clean and cheap source of power has faded beneath the surge of negative public opinion, and nuclear power no longer seems a viable option for consumer energy in this country. It's remarkable to think that clean and cheap energy in fission, one of the greatest promises of technology only fifty years ago, is to all intents and purposes now dead in the water.

It was the growing sophistication with which scientists, legislators, consumer advocates, environmentalists, and bureaucrats addressed individual risks to health and safety that resulted in the intrusion of risk-management devices into every corner of our lives over the last quarter century. All foods and beverages now bear nutritional labels, and some carry health warnings. Public health information pulses into our heads frequently, reminding us with warning labels, signs, buzzers, or lights to use a condom during sex, buckle up in the car, wash our hands before eating. Other devices involuntarily keep us from hurting ourselves or others: crosswalks, speed bumps, no-smoking laws inside buildings, the concept of a "designated driver." These intrude into the very fabric of life in ways that would have been unimaginable to our grandparents just two generations ago.

So substantial are these antirisk devices that they've fundamentally altered our relationship to risk itself. No longer can people so often look to blind fate, blame the whims of a deity, or just shrug their shoulders when something bad happens. We simply know too much about how the world works—basic cause and effect—to rationalize loss or harm in these ways. The accidental poisoning of a child may still seem attributable to the work of a higher power, but it doesn't prevent us from understanding that a safety cap on the medicine or a locked cabinet might have prevented the tragedy.

To complicate matters further, much of the confusion in modern America stems from who's responsible when the loss or harm of a source of risk manifests itself. For by digging deeply into risk, it's possible to apportion blame and assign responsibility. But to whom? The individual, a company, the government, another person or group of people? The fear of crippling personal injury lawsuits prompts manufacturers to take extreme measures. One day, when I went into a store to buy a Batman outfit for my son. I found a warning label on the costume, which read: *FOR PLAY ONLY: Mask and chest plate are not protective: Cape does not enable user to fly.* My son wanted to know why I was laughing so hard. As I walked out of the store, I tried to remember if the real Batman character could fly anyway.

The public carries its share of responsibility for the fact that corporations publish such warnings: consider the woman who sued McDonald's when hot coffee spilled in her lap, as if the quality of "hotness" was not an inherent characteristic of the beverage. Even if McDonald's did heat its coffee to higher temperatures than most restaurants do, it's still difficult to understand why it should be responsible for the woman's error of dumping hot coffee in her lap. Thus, ladders warn about the risk of falling, even though the exact purpose of ladders is to raise a person above the ground. These ubiquitous warnings come from a sensibility that the manufacturers of a product, such as a ladder or coffee, are liable for damages if they fail to warn consumers about possible risks. Defining and delineating matters of risk confront us with a difficult question: where does consumer responsibility end and manufacturer's negligence begin?

Perhaps the most disturbing aspect of rampant risk-management devices in our lives is the way they can begin to shape our sense of responsibility—or lack thereof. What individual needs to evaluate the riskiness of an action, if everywhere we turn a risk manager has instituted safety precautions? Ceding away the responsibility to make our own determinations of risk can make us a nation of victims, looking to blame everyone but ourselves when something bad happens. If a person walks in a state park with high cliffs, he can almost always count on a guardrail, fence, or sign to warn him about the danger of falling.

In modern American society it thus becomes easier and easier to rely on others (experts, regulators, bureaucrats) to look out for personal safety issues that are surely matters of common sense. Of course agencies, commissions, and experts should continue to evaluate the safety, health, and environmental risks of technology and new products. It is their encroachment into the realm of common sense that's disturbing. With ubiquitous risk-management devices in place, we no longer have to exercise our common sense: we can wander through life without paying attention, except when signs or other risk devices warn us of possible danger. How close can we get to the train on the safety platform? A yellow line or row of blinking lights will tell us. The implicit assumption in many cases today is that the lack of warnings and risk-management devices means that a particular activity is safe or risk-free.

Issues of risk and responsibility force us as individuals and society to begin drawing lines. Consider again that state park with high cliffs. Does a single warning sign near the cliff suffice, or should a warning also be printed and clearly visible on the park literature handed out to each visitor? Should there be a fence along the cliff edge? A fence represents a big ideological as well as physical leap from a sign: a sign gives us information — arguably a good thing even if the message is self-evident — and we can then choose what to do with that information. On the other hand, a fence over the cliff is an involuntary device that physically prevents us from walking close to the edge.

But what if this risk-management device obscures the view? Perhaps there should be no fence, but a rule that children under a certain age may not approach the edge. Maybe no one should be allowed to approach the edge except in the presence of a park guide, but that would entail an unacceptable loss of liberty. Inherent in this simple example of the difficulty of determining how to exercise risk management are issues that apply to millions of decisions that we must collectively make. At what point does personal freedom become more important than safety? When do involuntary safety devices make sense, and when do they become invasive, offensive, unnecessary?

The only way to make headway against these difficult questions is for all of us to become good managers of risk. However, even knowing the pitfalls, difficulties, and complexities inherent in managing risk doesn't equip us to become effective risk managers.

CHAPTER FIVE

POLAR BEAR FOLLIES, OR HOW UNCONSCIOUS RISK RULES CAUSE TROUBLE

Tempers had grown quite short, the reader will recall from the preface to this book, after our small canoe expedition to the Arctic had discovered signs of polar bear activity. The discovery had unnerved all of us, because out on that barren landscape there was no place to hide—there were no trees to climb, nor were there any safe havens in the event that we did surprise a bear. Worst of all, as noted, none of us could agree on a course of action, a strategy to deal with the risk. The disagreement reached a head after dinner on the day when we discovered that a polar bear had unearthed a warren of small mammals on the banks of the Kuujjua River, which we had spent the better part of two months paddling.

As far as the rifle we carried for protection went, except for one man, we were the gang that couldn't shoot straight. We counted among our midst a nuclear physicist of some renown, an international banker, the headmaster of a secondary school, and a business consultant. All of us were successful, white-collar professionals, used to attacking problems, analyzing risk, and coming up with solutions.

"Somebody should stand guard," suggested the nervous physicist. "Maybe the gun should reside with the guy who is the best shot," offered the banker. "No," someone else said, as the conversation seesawed back and forth, "the gun should be outside in a communal spot." I joked that we shouldn't shoot a bear at all, he'd just get pissed off. That didn't get much of a smile from anyone. The conversation started getting personal—not a good sign, since finishing the trip depended on our teamwork.

Finally, the management consultant (and incidentally, the only reliable shot in the group), who had some experience with gatherings that go astray, told a joke. He mentioned that we didn't need to worry, we just each had to have a good pair of sneakers and be able to run faster than at least one other member of the team. That drew chuckles, but I noticed that each of us surveyed the others slyly, suddenly thinking about them in a new light. I can run faster than *he* can, I thought, looking at the physicist. Curiously, after that joke, the animosity died, and the question of a polar bear defense never came up again. The gun remained in one of the packs. It seems the joke had a ring of truth to it: because everybody thought themselves faster than someone else, nobody worried anymore.

The element of truth in this anecdote is that there seem to be certain common denominators in any group's response toward a perceived risk that transcend differences in age, experience, and general disposition toward risk. In fact, an entire subdiscipline of risk analysis science—risk communication—is devoted to the idea that a raft of hidden, informal rules and assumptions colors human perceptions of risk. The "communication" element refers to how people perceive risk and how it can be explained or communicated. These general theories are loosely combined into a series of rules and assumptions, known as "mental models" and "intuitive theories," which an individual unconsciously applies to a range of risk decisions, from the mundane (crossing at a crosswalk or not) to the serious (whether to undergo dangerous surgery). How we perceive the likelihood of a risk, or its threat to us, is determined by a clutch of general theories about ourselves and our abilities, assumptions, and innate biases. These act as general guideposts that let us negotiate the nearly incalculable number of risk decisions we make every day. Considering each and every decision would slow us down to a crawl, so we rely on general rules of thumb about risk to guide us.

Learning these hidden biases and theories is akin to recognizing how a particular family member pushes your buttons, inciting you to anger, guilt, or sadness. Once you learn these buttons and how they're pushed, you can neutralize your feelings. As a result, family occasions become more bearable. In the same way, preconditioned risk responses and sensibilities can lead us off track, interfering with our ability to make sensible decisions that achieve the desired consequences. A key aspect of the modern language of risk is learning how our unconscious preconceptions influence our evaluation of the risks around us.

One of these general theories, the notion of "optimistic bias," was at play in our Arctic camp. Psychologists have noted that in many situations individuals tend to believe their skills are above average. For example, optimistic bias commonly reveals itself in polls on driving habits. When asked to rate their driving skills, most people will respond that they're above average. By definition, of course, only 49.9 percent of drivers can be above average. Consequently, a substantial number of drivers believe themselves to be at less risk while driving than they actually are, if one assumes that possessing above-average driving skills indeed reduces the chances of having an accident. Neil Weinstein of Rutgers University has accumulated convincing data suggesting that the optimistic bias comes not so much from overrating one's own capabilities but rather from underestimating those of others. In relation to driving, therefore, many drivers believe that they're above average not because they consider their own skills stellar, but because they view other drivers as worse than they actually are.

The idea that defanged the argument in the Arctic camp played into this notion of optimistic bias: "I can run faster than at least one other person here, so I needn't worry about the risk, because other people are slower than I am." Clearly, as in the driving example, not everybody in camp could have been right, even though everybody thought he was. This psychological predisposition of optimistic bias can hinder our ability to make accurate risk assessments. In the case of driving, optimistic bias skews many drivers into minimizing the true hazards of driving. The overconfident driver may drive fast in inclement weather, for instance, feeling safe because of an inflated sense of his own skills. In the Arctic, the optimistic bias, along with other factors, clearly helped to satisfy (falsely) each individual's comfort level.

Risk communication is fascinating because many modern risks and hazards, like the polar bear we might have faced, are removed from direct observation. With the sweeping changes in the risk landscape and the shape of technology, individuals are called on to judge the risks of developing chronic diseases that take decades to emerge, the impact of toxic chemicals that are invisible, and the implications of other environmental issues of great complexity. The discipline of risk communication received a big boost in 1980 with the publication of a groundbreaking study by psychologists Paul Slovic, Baruch Fischhoff, and Sarah Lichtenstein. They asked a diverse sampling of the population to rank thirty activities and technologies by level of risk and then compared the results with the rankings assigned by a panel of risk-assessment experts. The two groups agreed in some instances, such as the risk of motor vehicles, placed number 1 by the experts and number 2 by the public. But in others, large discrepancies appeared. The public rated nuclear power as the number 1 risk, whereas the experts ranked it a lowly number 20. Experts ranked x-rays number 7, while the man in the street put them at number 22. The study revealed incontrovertible evidence of the gulf between the risk perceptions of the public and those of the experts. The risk researchers concluded that the experts and the public disagree on the nature of risk because they use different criteria to judge risk. Experts use mortality as the sole guide to whether a particular event, industry, or activity involved risk to a person. In contrast, the public evaluates risk through a complex web of theories and assumptions.

In the arsenal of risk assumptions, individuals regularly assign a higher likelihood to situations that are "disproportionately visible" or easy to imagine. In other words, if an individual sees something frequently, he or she tends to assume it has a high rate of frequency. Herein lies a major reason why a person's risk perception can become skewed. A major source of information, "the disproportionately visible or easy to imagine," is the media, particularly the nightly local television news. What's most visible happens to be what's in the news and what people see in their own lives. The vision of the world presented by the local television news—full of violence, natural catastrophe, illness, and oddity—is familiar to all of us. What's news by definition is not the common and mundane but the unusual, exotic, and strange. From reading only front-page stories and local news headlines, one would imagine cities to be war zones and the Midwest to be buffeted constantly by twisters. Often the media do not present these glimpses of the exotic in any statistical or probabilistic context, so viewers are not given the information they need to gauge the frequency of these occurrences, and the consequent risks to them, by the media's attention to them. High-profile coverage of crime can thus produce an exaggerated sense of the prevalence of crime, even when overall incidents are falling.

In a similar vein, a case of deadly botulism that comes from a can of soup and kills a family will receive far more play than the crash of a minivan that kills another. A single event will saturate television news and news magazines, news and talk radio, newspapers, and print magazines. The oddness and horror of the botulism incident assure its much more widespread coverage in the media than the van crash, even if the number of deaths is the same in both cases. Of course, no one would ever believe that botulism kills more people than highway accidents; they would tend, however, to overrate the risk of botulism poisoning following such widespread reportage of an incident. Likewise, a single victim of flesh-eating bacteria or a consumer-product tampering will lead Americans to overestimate their own probability of experiencing these events. I am not arguing that a person shouldn't watch or read the news — it is the source of new information the public needs. Even new, small risks can, and should be, minimized. The key point remains that the way in which the information is presented can often skew our estimations of risk frequency and severity, and that wrong estimations of risk can lead to bad risk decisions.

Exotic risks also appeal to another hidden bias that unconsciously colors human perception of risk. Risk communication experts label this "dread." We all dread sudden, bloody, and violent events that can change the course of our lives in a few moments. Dread is heightened by the exotic — things we don't know about or can't see: the ravages of flesh-eating bacteria or the lethal poison of botulism in a can of soup. Death by shark attack or disintegration in an airplane crash is therefore a more appalling prospect than death by stroke, even though the end result is the same. Catastrophic events stick indelibly in our minds. Risk experts have discovered that people regularly overestimate the occurrence of dreadful risks and underestimate the odds of less-dreadful ones. The appalling event takes up more of our attention, even though a much more mundane event could prove just as deadly.

Dread appeared to play a major role in another risk study Slovic and Fischhoff conducted. They asked a representative sampling of the public whether more deaths in America were caused by accidents or by disease. Most responded that accidents killed more Americans. In fact, disease takes fifteen times as many lives as accidents do. Yet it is the specter of sudden, unpreventable, and dreadful accidents that captures our imagination and causes us seriously to overestimate the probability that they will befall us.

If a hazard appears not only dreadful but also unfair or inequitable, another factor affects our mental models of risk: "outrage." For instance, if an action appears to benefit one group while hurting another, then that risk becomes, in the eyes of the public, magnified. If a powerful manufacturing company dumps toxic waste into a site bordering a poor community, the risks of leaching toxins appear great to that community as well as others, even if the true danger is minuscule. A sense of outrage kicks in because the community is the recipient of hazards and no benefits, the rich taking advantage of the powerlessness of the poor.

Outrage also results when trust is broken, as it was when the Jensen Beach, Florida, dentist David Acer infected as many as six patients with HIV. The image and voice of twenty-three-year-old Kimberly Bergalis on her deathbed, calling for HIV screening of health care professionals, is something few can forget. How does this kind of outrage translate into skewed perceptions of risk? This one event made many people fear dentists, even though scientists at the Centers for Disease Control and Prevention in Atlanta calculated the chances of getting HIV from a dentist at 1 in 2.6 million, similar to the lifetime risk of being hit by an extraterrestrial object and less than the average lifetime chances of dying from accidentally inhaling a toothpick after dinner. Given the widespread reporting of the dentist story, I'm certain that it was hard for many people visiting a dentist soon thereafter to banish eerie thoughts. Now all dentists wear rubber gloves, although this precaution is taken more to protect themselves.

The way we perceive risk can predispose us toward subtle errors in judgment and logic. For instance, some people believe that natural is better than artificial, regardless of the circumstance. A parent, thinking that herbal remedies are superior to conventional medicine, might avoid medicating a child or having him vaccinated against for whooping cough. The result could be fatal sickness or an outbreak that puts other people at risk. On the surface, buying bottled water may appear to promote good health because a person can avoid the chlorine and other chemicals found in municipal tap water. However, since the manufacture of bottled water is largely unregulated, some types may contain harmful minerals, impurities, or micro-organisms not killed by chlorine.

The belief that bottled water presents less risk than city water plays into another general intuitive theory that hampers our ability to evaluate risk accurately. We automatically overestimate the risks of products or substances that are artificial and underestimate those of natural ones. As we'll see in later chapters, the chemicals found in many vegetables, although natural, can pack the same carcinogenic punch as artificially made chemicals such as pesticides. Yet individuals tend to assign human-made products a greater potential for harm. Radon, a naturally occurring radioactive gas that seeps into many basements, doesn't seem as sinister as radioactive emissions from a nuclear power plant, although its impact on human health is far worse. Many Americans routinely expose themselves to large doses of ultraviolet light while sunbathing yet become squeamish about minuscule or nonexistent radioactive leakage from a nuclear power plant or the radiation from a chest x-ray. Each year, several people die from drinking teas they brew from herbs they find in nature, victims of a fatal misconception that natural can't be harmful.

Of course, another general intuitive theory comes into play here: individuals often distinguish between risks incurred during voluntary activities or exposures and involuntary ones. Another bias that skews a person's perception of risk is the notion that she can tolerate much higher levels of risk from an activity, product, or exposure of her own choosing. Thus, an individual can ski downhill and drink alcohol all afternoon—both high-risk activities—and still become concerned or outraged about the preservatives in her dinner that night—a comparatively low physical risk. To be fair, some people who ski and also worry about pesticides see risks and benefits in these activities besides health-related ones. A person, for instance, for whom skiing is a passion may consider the sport's physical risk a small price to pay. But a person who sees the residue on a lettuce as part of a larger pattern that fattens growers' and wholesalers' profits while poisoning workers and the environment cares about a range of different issues, not just his or her own health protection. It's safe to say that regardless of the absolute level of risk involved, whether an activity is voluntary or not can sway how we evaluate its associated risks.

In extreme situations, advocacy groups and advertisers can manipulate public fears to advance an agenda by presenting a risk in a manner that plays upon the risk theories or biases that magnify the perception of risk. A prime example of this occurred during the Alar scare of the late 1980s. Eager to publicize their concerns about agricultural chemicals, an environmental group, the National Resources Defense Council, fed reporters at CBS's *60 Minutes* a story that daminozide (Alar is the trade name) put children at great risk for cancer. Used by a small percentage of the nation's apple growers in the 1980s, mostly on the red delicious variety of apples, Alar is a growth-regulating chemical used to keep apples from falling off the tree too early. With the help of a shrewd public relations consultant, the environmental group pushed the public's risk buttons through the media.

60 Minutes correspondent Ed Bradley, sitting on a set featuring a large apple and a skull and crossbones, reported that Alar was the "most potent cancer-causing agent in the food supply today." During the report, many aspects of the public's mental models and general intuitive theories about risk were inflamed: dread, because an invisible substance allegedly was killing children; outrage, because children were allegedly being poisoned for profit by a large company; natural versus unnatural — the juxtaposition of the symbol of nature and health with an artificial chemical; voluntariness, because many kids had no choice but to eat Alar-tainted apples if they wanted to eat apples.

From the public relations perspective of the environmental group, the report proved a big success: within twenty-four hours, the country erupted in fear and panic. A woman called the Environmental Protection Agency and asked if she could pour her apple juice down the drain or whether she should take it to the toxic waste dump. School districts in Los Angeles, New York, and other cities immediately banned apples and apple products from their cafeterias, and many millions of dollars' worth of apples were dumped into ditches. Actress Meryl Streep, a concerned mother, joined the fracas: in a television commercial she used detergent to wash pesticides off vegetables; she also testified to Congress. The public appeared to believe that one bite of an apple treated with Alar could strike you dead. The apple industry lost more than $100 million, and a number of small-scale growers, many of whom had never even used Alar, went out of business.

The panic originated from a controversial study of questionable validity dating from 1973 in which laboratory mice developed tumors when exposed to 35,000 times the amount of Alar that children were normally exposed to. After the uproar caused by the media report, Uniroyal, the maker of Alar, took it off the market. Later, several independent reviews found Alar's threat to be minuscule, but by that time the notion of "deadly Alar" was firmly entrenched in the public mind. While in retrospect the public reaction seems overblown, concerns at the time were not unreasonable. Generalizing from moderately accurate beliefs about the danger of many agricultural chemicals to one that is unknown is reasonable.

Certainly, trusting *60 Minutes* and a hitherto reliable environmental group was reasonable as well. Given little concrete information or any context for evaluating the risk of Alar, members of the public were forced to come to their own conclusions. The fact that this issue was presented in a certain way inflamed many unconscious risk biases and led people to misinterpret the true risk.

Few instances smack of such clear public manipulation. More often, there's no organization intentionally attempting to delude the public. In many cases, for instance, a study is published that establishes a preliminary link between an activity, substance, or product and a health risk. If the risk presses the right "hot buttons," many people grow very concerned, or even hysterical, precipitately overrating the danger and adjusting their behaviors. Even should the preliminary findings prove false and the risks negligible, the sense that an activity or product is dangerous can linger. This phenomenon occurred with electric blankets, the scare over coffee causing pancreatic cancer, cellular phones and other electromagnetic field–producing systems, and asbestos in schools and workplaces. Often there's a gap of years, or even decades, between the time when a new risk is first suggested and the time when scientists can determine whether that risk is minor, severe, or phantom. It's not unreasonable for the public to grow concerned about a newly postulated risk that may prove to be dangerous and worth aggressively addressing. However, a prerequisite of speaking the language of risk is realizing that certain kinds of risk cause strong responses, which can sometimes lead to overreaction.

It's difficult to know how much perceived risk is needed to push a person to act, as the mother did who poured apple juice down the drain because of the Alar scare. While the general theories, biases, and assumptions about risk may well be universal, individual risk tolerances vary greatly. In the Arctic, for instance, some of our group felt extremely nervous about the polar bear threat, their sense of dread and concern causing them great anxiety. Despite such individual differences as these, a provocative and highly controversial idea suggests that while individual risk tolerances vary, each individual keeps a mental running tally of risk, constantly adjusting risk behavior in order to maintain an overall risk balance.

According to Canadian psychologist Gerald Wilde, the key to this new theory lies in the term *homeostasis*, which he borrowed from medicine and physiology. Wilde argues that individuals constantly alter their behavior according to a perceived threshold of overall risk. When they feel that their overall risk has lessened in one area, they feel they can take greater risks in another. Risk homeostasis is at play, Wilde suggests, when a person drives faster than he would normally when he gets behind the wheel of a car that is larger and safer than his own. The extra safety factor of the large car enables him to take greater driving risks yet still maintain the same overall degree of safety he enjoyed in his regular car. Similarly, a woman who has a gun in her pocket might go to places she wouldn't have gone otherwise, confident that the gun buys her greater security.

For many years, physicians have known that the body maintains a certain equilibrium in its life support systems, such as core temperature, heartbeat, blood pressure, and blood sugar. Unless such basic functions were kept within certain boundaries, our bodies would die in short order. In the case of blood pressure in our arteries, for instance, the heart pumps oxygen-rich blood out into the body through a series of arteries that gradually diminish in size. Around the smaller, more distant arteries are muscles that can contract to increase the blood pressure and delivery of oxygen to tissue. If the brain perceives immediate danger, it can signal increased blood pressure by telling the heart to beat faster and the muscles around the arteries to contract. As a person goes to sleep, the brain lowers the body's blood pressure. Homeostasis doesn't mean maintaining constancy, as a thermostat does in the home; it is, rather, a dynamic process that accommodates the shifting needs of the body. Wilde believes that individuals keep a running tally of risks in the back of their heads and keep adjusting it as they need to, depending on feedback and new information.

Wilde's theory builds on the more traditional risk-taking theory of "expected utility maximization, found in economics and decision-making theory. This theory posits that an individual will incur risks in proportion to the benefits the person expects to reap as a result of taking those risks. In driving, for instance, no one can expect zero risk of an accident. That would mean never driving at all. One could avoid a fatal driving accident by driving a tank, but the inconvenience of getting places (not to mention the cost) obviously makes this impractical. At the opposite extreme, everyone could drive motorcycles at excessive speeds, reducing travel time in cities significantly but becoming much more vulnerable to accidents. Most of us pick something in between, selecting a level of risk we are willing to tolerate based on what we feel is convenient, appropriate, safe, and cost-effective. "The human being," writes Wilde, "is seen as a strategist, a planner, who attempts to optimize, not minimize, the level of risk-taking for the purpose of maximizing the benefits—economic, biological, and psychological—that may be derived from life." Wilde argues that individuals maintain target levels of risk in different areas, such as driving. Add all the safety devices available, suggests Wilde, and people will nevertheless maintain the same general level of riskiness in their behavior.

To bolster his theory, Wilde cites a German Ministry of Transport study of Munich cab drivers. When part of the city's fleet was updated, the new cabs were equipped with an antilock braking system (ABS). By preventing the wheels from locking when the driver puts on the brakes quickly, ABS gives the driver more control, especially on wet and slippery roads. But after the German authorities secretly monitored the entire fleet for three years, surprisingly, the cabs with ABS showed a greater frequency of accidents despite the safer braking system. Incredibly, the addition of ABS appeared to make cabbies less safe because those drivers drove faster, made sharper turns, had more near misses, and braked harder than cab drivers without ABS. Wilde surmises that the cab drivers with ABS changed their driving behavior, believing that they could drive more aggressively with less risk than before. In essence, they used up the increased safety margin provided by ABS to drive with more abandon.

A similar situation occurred in Sweden during the 1960s, when the country switched from driving on the left side of the road to driving on the right. Instead of increasing while drivers learned the new system, traffic fatalities actually dropped 17 percent over the following year. Aware of the dangers of their newly unfamiliar driving conditions, Swedish drivers compensated by driving more carefully. Eventually, accident levels reverted to their previous levels. Wilde believes that these examples support his idea that risk-management strategies must motivate people to drive more safely. Making a curve on a road appear tighter than it really is so that people slow down more readily could, ironically, be more effective in saving lives than signs or even guardrails.

The risk homeostasis theory nevertheless has significant problems, not the least of which is that it is difficult, if not impossible, to prove empirically, making it next to worthless in the eyes of most scientists. Perhaps its greatest difficulty is that it doesn't deal adequately with the nonrisk factors that often play an important role in how individuals make decisions and view risk. Americans, for instance, could be taking much longer driving vacations, and putting themselves at greater risk, not because they view their risks any differently but because gasoline prices have dropped substantially. Still, the possibility that individuals adjust their behavior in an attempt to keep their overall risk constant remains an intriguing idea.

What is indisputable is that individuals do seem to have different tolerances for risk. This reality was brought clearly into focus for me on a recent trip I took to Norway with my wife. Not far from the city of Loen is Ramnefjell, a remote fjord that our guide insisted we visit. This fjord was especially pretty, although curiously empty of human habitation. A strip of lush tree and bush cover clung to the foot of the cliffs a hundred feet above the water, wide enough for a single-lane road and an occasional house. At the water's edge lay an even lusher flat expanse of rich bottom land. In Norway, where due to the extreme topography only 3 percent of the land is arable, this land seemed as though it ought to be prime real estate. So hard pressed were they for land that Norwegians of previous centuries scraped by on subsistence farming on the edge of sheer cliffs: many families were known to tether their children and livestock to prevent them from a misstep that might send them to their deaths. Here in this pastoral fjord, no such danger was apparent. Why was this land uninhabited?

When we reached a particularly spectacular section several miles in, our guide stopped. Across the fjord lay Ramnefjell Mountain, an immense chunk of sheer basaltic rock, rough-hewn and dark with age. Below us on the land at water's edge, the rusty red hull of the steamer *Loden* lay high and dry, upside down. As the guide told us the story, the fjord revealed its murderous aspect. Scores of people had once lived here on small farms. Then, in January 1905, a monumental slab of this wall split off and crashed into the fjord below, generating a giant wave later estimated to be about 225 feet high.

Back then, the Bødal and Nesdal families lived here, probably engaged in dairy- and other farming. The wall of water crashed across the fjord with a force hardly imaginable. With nowhere else to go, the wave tore through the narrow fjord, dissipating only after it claimed sixty-one lives. Those very few who escaped by climbing above the water seemed to remember a deep rumbling sound above all else. A memorial plaque riveted into the cliff here lists the names of those who died in this awful tragedy.

No strangers to tragedy and hard lives, the survivors returned to their farms and rebuilt, living and working under the same mountain that had killed many of their family members. Perhaps it's optimism, but something deep inside us argues against the likelihood of tragedies striking twice in a row. When violent acts of nature occur fiercely and without mercy—whether in a tornado, hurricane, or tsunami—we have difficulty imagining it happening again soon in the same area. It's as though we ascribe a sense of equity to nature. We all know that lightning never strikes twice in the same place.

Incomprehensibly, though, disaster did strike Ramnefjell again in September 1936, when another piece of the wall sheared off, sending another immense wave raging through the fjord. This time seventy-four people lost their lives, all of them intimately acquainted with the first disaster, and many descendants of those who had died a generation earlier. It's not hard to imagine a young man who had lost his grandfather to the first wave often gazing up at the cliff and wondering how it had all unfolded—only to find out for himself when the second wave swept the fjord. A new list of victims joined the old on that plaque, many bearing the same last names.

As we left, I saw a lovely wooden cabin for rent and stopped our guide to take a look. I peered through the windows and imagined a summer's season there enjoying the grand unspoiled views of Ramnefjell Mountain. Back in the car, my wife met my dreams of distant summer vacation with a look of amazement, aghast that I would even contemplate the idea. I answered, "Don't be silly, it won't happen here again." I realized then that my wife and I had looked at the same set of facts and come up with two entirely different views of the risks involved. The history of this fjord caused her to fixate on the chance of its recurrence. I, in contrast, considered the likelihood to be negligible. After all, what would be the odds of a third disaster occurring during the one week we might spend here on vacation? Still, I recognized that I wasn't cooly evaluating the risks. I had no information about the geology or weathering process of the rock wall, just the belief that because a tragedy had happened twice already one wouldn't happen again soon. I had no evidence, just a gut sense.

When I considered that cabin in Norway as a future vacation site, I saw one thing, while my wife saw another. She imagined our children and a roaring wave of water. On that basis, she probably exaggerated the probability of such an event occurring. On the other hand, I probably underestimated it. Our respective perceptions of the risk involved had little to do with the empirical evidence. And we each arrived at entirely different risk estimations. Lacking any evidence, we brought strong intuitive theories to bear to gauge the extent of risk involved. My wife, of course, won out. We would never rent that cottage.

CHAPTER SIX

I'M RISK PRONE, YOU'RE RISK AVERSE

New evidence in a field outside social science suggests that my wife and I, the other members of my expedition to the Arctic, and individuals generally are influenced in risk estimation not only by psychological factors but by biological factors as well. On a trip out to northern Utah one fall, I happened upon a group of young rock climbers battling a sheer cliff face in the Cache National Forest north of Logan. The vertical limestone cliffs there draw a clique of high school boys every afternoon. The river has cut a deep trench in an ancient sea bottom. On that weekday midafternoon, fluorescent nylon ropes hung on the wall, flipping and wriggling like the tails of nervous lizards. A half-dozen clumps of young men huddled near the rope ends, casting wary eyes above to climbers stuck like spiders on the wall face.

There wasn't much laughter or high jinks that I could see, just a scattering of shouts encouraging someone beyond a tough overhang. I stood there mesmerized. After about an hour, the energy of the group suddenly changed and all eyes cast over to a section of rope where one young man had rappelled down from the top to a spot about 30 feet from the ground. It was as though a beautiful young woman had just entered a locker room packed with adolescent boys; an atmosphere of desire, focus, and nervousness spread among the boys along the cliff bottom. "He's crazy," I heard one young man mutter. "Wow!" said another. Above us, the curly haired young man kept bouncing out into the air swinging on the end of his harness, with a serious expression on his face. Then in one moment, at the end of a swing, he released his braking grip on the rope and plummeted nose first in a swan dive earthwards with heart-stopping velocity. Instants before his body would have hit the earth in a fatal impact, he arrested himself with a quick braking tug on his rope and bounced to a stop. His body jerked like a puppet on a string, but the rope above held fast. The young man dangled just a few feet above the ground for a few heartbeats, righted himself, and then let several others help him unclip from his harness. I saw him quickly flash a smile, and then all the groups went back to their own climbing. I overheard someone call this "Australian rapelling."

Young men aged sixteen to nineteen have always taken more risks than any other group, a result of rampant hormones, youthful inexperience, and extreme desire to experience thrills. Even in this group, however, there seemed to be limits to what some would do. I felt certain after watching their faces that most enjoyed the thrill of climbing, an activity that involves considerable risk. While all of them rappelled, however, clearly few in this crowd would ever consider rappelling face down in such a bold manner. Perhaps the curly headed young man derived high standing among his peers, but the pure brazenness of the act spoke of something even greater than a ploy for prestige or status. The act seemed to satisfy him powerfully on a physical level. In the act of rappelling, this young man had a different view of personal risk than his companions did.

Back at home, wrapped up in the more mundane matters of everyday life, I began to pay attention to differences in how different individuals aggressively court or avoid risks. Life offers incalculable opportunities for risk taking: going out on a date, driving faster than the speed limit, confronting a boss. Of course, few everyday activities involve life-threatening experiences or high adrenalin rushes. Nonetheless, taking risks, even on a day-to-day basis, can involve feelings of satisfaction, even euphoria, for some people and dread for others. I began to sort people out according to their particular relationship to risk. Some friends relish the chance to embrace a risky venture, whether it's betting on the spread of a football game or sinking all their retirement investments into highly volatile technology stocks. Other friends wouldn't dream of betting and prefer conservative investments in government bonds. They place a high premium on minimizing risks in their lives.

One person's high risk is another's lark. Anyone who's ever managed people, maintained friendships, or kept a marriage healthy has a good working knowledge of this basic tenet. It's a fact of life. Some people seek great physical risks as war correspondents or professional mountain climbers. Others avoid them at all costs. Some academics seek high intellectual risks, others shun them. Until recently, psychologists believed that much of these personal differences among people were rooted in socialized or learned behavior. New research suggests that some of our predisposition or antagonism toward risk is rooted in biological functions. In other words, part of the reason a person is risk averse or risk prone may be hard-wired, that is, built directly into the individual's physical makeup. Over the last several decades, biochemists and psychologists have assembled theories about personality, gender, and built-in response to risk in an attempt to explain these differences among people. Recent discoveries in genetics have engulfed the discussion in controversy. The debate opens up fascinating insights into why some of us are risk takers and some are not. Our bodies, in addition to psychological and environmental factors, may predispose us toward regarding risk in different ways than our fellows.

All humans share a common response to perceived risk when it's immediate and potentially life-threatening. You have probably felt your heart pound furiously, your breath deepen, sweat beads appear on your brow, and your mouth grow dry. These automatic reactions are part of the nervous system, and individuals have little control over them. In a stressful situation, the body undergoes a quick, rather monumental change in its biochemistry: neurotransmitters cause the synapses in the brain to fire more quickly; the endocrine system causes muscles to tighten; blood pressure rises. Even the blood chemistry changes. The amount of blood sugars, antibodies, and proteins increase. Without such protective mechanisms, *Homo sapiens* as a species would never have survived. When one of our early hominoid ancestors suddenly faced a saber-toothed cat, he needed all his physical resources immediately. Deep breathing enriched his blood with oxygen, and a faster heart rate spread that oxygen with optimal speed to muscles that might be called upon to run or climb a tree.

No one, of course, enjoys the flat-out feeling of gut-churning fear. However, slightly stressful situations, such as whipping a motorcycle along a country road at high speed, can lead to exhilaration. Between sheer terror and normalcy lies a wide spectrum. Some people seem to crave feelings associated more with the terror side, whereas others prefer rarely to press beyond normalcy. Some individuals so crave physical risks that they make a career out of it. It's why some journalists jump from one war zone to the next, whereas others are content to cover local beats. While the derring-do of the young man performing an Australian rappel could be explained by his age and impulsiveness, or his inability to realize the seriousness of the consequences, the motivation of a committed, mature adventurer who keeps coming back for more is more difficult to explain.

Sheck Exley, a cave diver celebrated in *Sports Illustrated* as the world's master of this dangerous sport, is a case in point. Cave diving is not for the faint of heart; stories of the hazards could have been penned by Edgar Allan Poe had self-contained underwater breathing apparatus (scuba) been invented in his day. Dark, often cold, and largely uncharted, underwater caves are one of Earth's last true frontiers, presenting a litany of dangers to those who enter these strange realms. Divers encounter ceilings that crumble overhead. Sometimes their fins disturb silt that settles and obscures exit ways. Equipment malfunctions in these harsh environments. Vertigo and disorientation can grip even experienced divers who enter chambers or swim down wrong passageways. Victims have been found several feet from silted-over exits, their air exhausted and fingers cut raw by scratching at the rock. Pure and simple panic claims others; bodies have sometimes been found with more than 30 minutes of air left.

In this world Sheck Exley, a high school calculus teacher from Live Oak, Florida, was a leader, having logged thousands of cave dives. Diving had consumed Exley ever since he was a young man. One day when Exley was nineteen and his brother Edward was sixteen, they gathered at Wakulla Springs in Florida. As young men are wont to do, they entered a competition, seeing who could swim deepest into the dark waters. Hyperventilating to enrich his blood with oxygen, Sheck dove, reaching 35 feet. Then, his brother went to 42 feet, each dive recorded on a wrist-mounted depth gauge. Again Sheck dove, pushing the record to 50 feet. Breathing furiously, Edward dove again. This time he never came up; he passed out and his body kept sinking. There was nothing that Sheck could do.

Despite this tragedy, Exley continued to dive, taking precautions and penning six books on cave-diving safety, including *Basic Cave Diving: A Blueprint for Survival*. When someone asked him about all this safety work, he responded that he didn't want his parents to lose their only remaining son. Exley became a legend, registering more than 3,000 cave dives. He participated in thirty-six body recoveries and captured the holy grail of diving, the depth record for diving, at 867 feet. Cave diving for depth adds even more dangers to the sport. Divers must mix their own tanks of helium and nitrogen. After 600 feet, high pressure nervous syndrome, or HPNS, can kick in. There's still little understanding of HPNS, a debilitating disease in which the immense pressure shrinks one's eyes and causes convulsions, hallucinations, and death, because so few have ventured so deep. When asked why he liked this sport, Exley answered simply, "I can't stop diving." Other adventurers who enter realms average humans can only imagine submit equally vague and frustrating answers about what drives their obsessions.

A few years ago, on a bid to recapture his depth record, lost to another diver, the forty-five-year-old Exley and friend Jim Bowden decided to try to crack the elusive 1,000-foot mark. Such dives require dozens of tanks, and the divers must spend as much as ten hours decompressing under water so as not to suffer a potentially fatal case of the "bends." They chose a sinkhole known as Zacaton in east central Mexico. On April 6, 1994, under the watchful eyes of a handful of journalists, the divers both descended on lines. Exley never came up. Three days later they found his body, tangled in his guideline, his depth gauge reading 904 feet. The legend was dead.

What does one make of such an obsession? By all accounts Exley was careful and an exacting professional. He shunned the spotlight, turning down spots on the morning talk shows. He had lost his brother and recovered dozens of bodies. Yet he came back to it time and again. Finally, something went wrong and he died. This kind of behavior in our society is difficult to fathom. What could possibly press someone to push boundaries and court death in this way? Perhaps Exley battled demons, his interior landscape tortured with things unknown to the rest of the world. But the evidence suggests not. A more convincing explanation lies in the idea that the risk of cave diving didn't loom as large to him as it does to someone else.

With forethought, he confronted the risks, taking precautions and relying on his own judgment. I've seen this in many mountain climbers and felt it myself. There's a powerful satisfaction in self-reliance and being prepared. Exley would have subscribed to the risk statement in a Federal Aviation Administration handbook: "The outcome of any maneuver must never seriously be in doubt." For him it wasn't. Belief in this axiom separates the daredevils from the serious adventurers. Yet the satisfaction of self-reliance still doesn't explain why people push the extreme, time and time again.

Some psychologists would have chalked Exley's behavior and ultimate demise up to death-wish fulfillment. This concept finds its roots in the work of Sigmund Freud, who no doubt would have viewed Exley's behavior as dysfunctional or abnormal. In Freud's view the organism is constantly engaged in reducing tension in its life. Voluntarily seeking an increased dose of stimulation, such as cave diving, is a matter of sublimation, displacement, attempt to deny fear, or a way of hiding inadequacies. Or perhaps Exley's risk-taking behavior was an attempt to build tensions so as to increase the enjoyment of reducing them.

New evidence suggests a simpler answer: some individuals may need higher levels of stimulation than others. In extreme cases, individuals may become addicted to the feelings associated with risk taking. Sheck Exley may have been one such person. Writes another adventurer, the British mountain climber Chris Bonnington, "The elation at being on a steep crag with the added spice of risk is something that has never left me. I reveled in climbing on loose rock with little protection." New findings in biophysiology suggest that risk taking may be the result of a biological function rather than an abnormal personality problem or psychosis. The shift from classifying extreme risk taking as a natural expression, rather than as a death wish, is revolutionary indeed and holds some fascinating implications.

As long ago as the fifth century BC, the ancient Greeks recognized individual differences as a matter of their physical makeup. A person's character or temperament consisted of a mix of four cardinal "humors," each related to a particular personality characteristic: blood (the sanguine), phlegm (the phlegmatic), yellow bile (the choleric), and black bile (the melancholic). The choleric man, for instance, was yellow-faced, lean, proud, shrewd, and quick to anger, whereas the sanguine man was cheerful and optimistic. A balanced person had equal representations of each humor, while all sorts of irregular behavior could be ascribed to an imbalance. Remarkably, the idea of humors in the abstract is not incompatible with what modern science has learned.

During his work on conditioning behavior in dogs during the 1920s, the Russian scientist Ivan Pavlov observed that some dogs fell asleep while harnessed and undergoing the often monotonous conditioning experiments. No doubt frustrated, because of course a sleeping dog is not responsive, Pavlov consequently took care to select the most active dogs. Returning to the lab, he observed something peculiar: it appeared that the sanguine dogs (the most active) tended on the whole to fall asleep more often than the melancholic (shy and less adventurous) ones. Pavlov speculated that these differences lay in basic differences in the dogs' respective nervous systems—some were easily bored and needed stimulation to stay awake, while the quieter dogs remained alert and awake through the conditioning experiments. Although Pavlov would become famous for his work on conditioning animals, his observations on individual differences would be forgotten for a while.

In the 1950s, British psychologist Hans Eysenck borrowed from Pavlov and others and developed a scale to measure individual differences in personality, isolating the trait of "extraversion." Eysenck wrote, "The typical extrovert is sociable, likes parties, has many friends. . . . He craves excitement, takes chances, often sticks his neck out, acts on the spur of the moment. . . ." In contrast, "the typical introvert is a quiet, retiring sort of person. . . . He tends to plan ahead, 'looks before he leaps,' and mistrusts the impulse of the moment. He does not like excitement. . . ." Esyenck posited that the difference between extroverts and introverts lay in the differing abilities of cortical neurons to respond to excitement. The introvert is happy with low levels of stimulation, while the extrovert would be bored and discontented with these same levels. The extrovert is a "sensation seeker," a person who goes out of his way to look for stimulation and excitement.

During the 1960s, Marvin Zuckerman of the University of Delaware refined Esyenck's work by focusing on the trait of "sensation seeking." He found that so-called high-sensation seekers tend toward participating in dangerous sports, volunteer for combat assignments in wartime, have an appetite for variations in sex and desire numerous partners, make bigger bets while gambling, smoke, take illegal drugs, and drive fast. Risk taking, in all its guises of physical, financial, legal, even mental risk, is closely linked to sensation-seeking behavior. However, high-sensation seekers don't seek out risk for its own sake; rather, they accept risks *in return* for the sensations: the rush of cocaine, the adrenalin of the high-stakes gambler rolling craps, the plummeting sensation of the bungee jumper. High-sensation seekers tend to be impulsive and gregarious. Conversely, low-sensation seekers do not value the sensations associated with risk as much as high-sensation seekers do.

All this research underscores the idea that people have consistent individual differences in their optimal levels of stimulation and arousal. But what's the root cause of these personality differences? Are they learned? Inherited from one's parents? Gender-specific? Controlled by biochemical reactions in our bodies? The nature/nurture debate remains one of the thornier problems in biology. Few would now argue that the human being is a blank slate, ready to be drawn on at birth by parents, teachers, and others. And few would argue that our genes and physiology entirely predestine our behavior and personality. Psychologists battle one another over the ground betwixt these two extremes. It is difficult ground precisely because it seeks to pinpoint the source of what makes us different from all other creatures. It is infuriatingly complex because each human being (with the exception of identical twins) carries a unique set of genes. In addition to these individual genes, each human passes through a wildly different social environment of home, school, friends, and so forth. Often, it's hard to separate the genetic from the learned — and the two forces can work either in concert or at odds. As psychologist Judy Dunn has pointed out, what about the fat boy with his equally obese parents? Is that boy obese because of a genetic predisposition passed on from his parents or because he lives in an environment where people eat a lot?

It is clear that the environment plays a large role in behavior. Psychologists use the word *environment* to describe any outside influences impacting the individual, from school and parents to television and peer groups. Studies of identical twins provide a strong case in point. Researchers have followed the lives of identical twins separated at birth. Since each twin shares the same genetic complement as his twin, if genetic influences accounted for entire shaping of personality, then such twins would be exactly the same despite their different homes. However, studies show that identical twins share only about 50 percent of their personality traits.

A recent study by Massachusetts Institute of Technology researcher Frank Sulloway advances a compelling argument for the influence of the environment. He argues that birth order substantially shapes personality and future behavior of individuals. In the past, many psychologists have cast the home as a single, unified environment. Sulloway believes that the experience of the family is quite different for every sibling as each vies for parental approval. He borrows the term "niche" from ecology to describe the different roles taken by each child. In the wild, organisms survive by carving out an existence defined by a specific set of conditions in behavior, habitat, and food. If another species invades their particular niche, their survival is on the line. In the family, firstborns typically align themselves with their parents, adopting a view of the family similar to their parents'. Firstborns then fight hard to protect this "niche" from later-born siblings, who, faced with an entirely different familial landscape, are forced to carve out a quite separate niche for themselves. Where the firstborns excel in conforming to parental expectations, later-borns, says Sulloway, "typically cultivate openness to experience." This strategy, he argues, propels later-borns toward taking greater risks.

Sulloway supports his thesis by analyzing the birth order of major Western scientists who have made revolutionary breakthroughs. Later-borns are responsible for most of these breakthroughs. In support of the status quo learned as children, firstborns often fight against radical new ideas. No one yet has looked to see whether high-sensation seekers are more likely to be later-born, but logic would suggest that they are. It bears repeating that these observations represent tendencies, not absolutes; nothing here is written in stone.

What do we know about the genetic and hereditary sources of personality traits? Within the last ten years, new gene-isolating technologies have put us closer to determining what is controlled by our DNA. Of course, certain physical traits, such as hair and eye color, are entirely determined by genes. The hair and eye color of identical twins (with, of course, identical genes) are always the same, suggesting that these traits are entirely genetic in origin. Other physical traits, such as height, are trickier to pin down. Using identical and fraternal twins, researchers have found that height is 80 percent inherited. Poor nutrition, for example, could influence the height of an individual. When it comes to less tangible traits, the heritable connection grows even murkier. Researchers have found that intelligence (as measured by IQ) is about 50 percent heritable. Even the most gifted individuals will have trouble developing that intelligence if they aren't given the tools, such as education, with which to express it.

In the realm of personality, inheritance plays even less of a role — between 20 and 40 percent, according to some researchers. High on the scale are such aspects of a personality as extraversion and neuroticism. It's important to remember that genes themselves are only chemical maps for the organism. Although they are powerful master plans, they don't deliver complex behavior patterns in nice neat packages. Encoded within DNA are mechanisms that build and control highly complex substances known as proteins. Proteins in their many guises are essential to life and the structure of the nervous system: as enzymes, they control the metabolism of the cell; as hormones, they control development; as neurotransmitters, they control how information, from pain to pleasure, is passed from one cell to another. Nothing in these chemical templates contains information that predestines an individual to become gregarious, quiet, or neurotic. However, an individual's genes do determine the type and regulation of proteins that make our brains more or less responsive to stimulation. Recent discoveries concerning the biochemistry of the brain yield some fascinating insights in this area.

In the brain, an important series of proteins known as neurotransmitters acts as the bridge keepers and toll collectors on the body's highway of nerves. Tiny electrical impulses shoot from nerve to nerve communicating between the brain and the rest of the body. Impulses communicate the sensation of cold in the fingertips, recognize danger and kick in a series of metabolic changes to help us survive. Neurotransmitters, such as serotonin, norepinephrine, and dopamine, are organic substances that regulate the nature and strength of these impulses. Like a world-class symphony, these different neurotransmitters act together to heighten and dampen responses. Serotonin, for instance, often acts to muffle the response of the brain to norepinephrine and inhibits arousal. The presence of serotonin can lead to the feeling of being relaxed. One of alcohol's pleasant effects is to release serotonin into the blood. Prozac does the same, helping ease depression. These intricate interactions make simple descriptions difficult, because the chemicals interact and act differently according to various situations.

The power of neurotransmitters was dramatically demonstrated by Oliver Sacks in his work with Parkinson's disease patients in the 1960s, a case popularized in the 1990 film *Awakenings*. During World War I, a disease causing inflammation of the brain, known as encephalitis, spread worldwide. Many of those who survived the disease developed Parkinson's disease, a degeneration of certain neurological pathways that leaves its victims catatonic, staring into space. In the 1960s, researchers discovered the precursor to dopamine, L-dopa. An ambitious young doctor, Oliver Sacks, administered L-dopa to twenty catatonic patients. Most of them dramatically "awoke." With their levels of dopamine artificially restored, the patients picked up where their lives had stopped abruptly in the 1920s. Without dopamine, their lives held no movement, joy, emotion. With it, miraculously, these were restored. Unfortunately, Parkinson's disease had destroyed the pathways' ability to function, and the patients slipped into bouts of manic depression. However, the seed was planted, and researchers grew fascinated with dopamine.

In early 1996, newspapers across the nation reported the stunning findings of two independent research organizations, discoveries perhaps more miraculous than the awakening Parkinson's patients. In two separate populations — one Israeli, one American — researchers found a genetic link between the personality trait of "novelty seeking" and a gene that controls how the brain responds to dopamine. People who rank high on the scale of novelty seeking tend to be extroverted, impulsive, and more inclined to take risks and search out new sensations. They also carry a slightly longer version of the dopamine D4 receptor gene than quieter, introverted people. This gene controls a receptor that is responsible for controlling how the brain responds to dopamine. This much-heralded finding represents the first time scientists have discovered the genetic source of a normal personality trait. As the *New York Times* reported, earlier studies of twins and laboratory animals had indicated that about 50 percent of novelty-seeking behavior is attributable to a person's genetic makeup. The authors of the 1996 studies believe that the D4 receptor gene accounts for about 10 percent of the novelty-seeking behavior among individuals. The researchers believe that other, as yet untested genes also contribute to a person's responsiveness to dopamine. Perhaps, the studies indicate, genes can influence the propensity of an individual toward or away from risk taking.

One of the principal tenets of the scientific method is that other researchers must be able to replicate the findings of an experiment. But less than a year after these results were published in the journal *Nature Genetics*, another study raised doubts. When American and Finnish researchers examined two groups of Finnish men, they found no correlation between novelty seeking and variations of the D4 receptor gene. These conflicting reports reveal the struggles of the new science of behavioral genetics. The original researchers were nonplussed. Said Dr. Richard P. Ebstein of the Herzog Memorial Hospital in an interview: "If you have six genes contributing to novelty-seeking, you're going to find that some people have all six genes in common, some have only three, some have even fewer. We're not going to be able to sort this out until we've identified all the genes involved."

University of Delaware researcher Marvin Zuckerman found another tantalizing clue to the biochemical connection between neurotransmitters and his scale ranking individuals on a sensation-seeking scale. (Sensation seeking describes many of the same features as novelty seeking.) In 1974 researchers discovered the existence of a substance known as monoamine oxidase (MAO), an enzyme that breaks down neurotransmitters and keeps them in balance in the brain. Later, Zuckerman found a correlation between levels of MAO in blood platelets and his scale of sensation seeking. High-sensation seekers tend to have lower levels of MAO than do low-sensation seekers. Zuckerman cautions that he's found only a "correlation," not "causation," between MAO and sensation seeking. In other words, no one knows yet whether low MAO levels actually cause people to be high-sensation seekers. Zuckerman has merely noticed a trend. There could be independent factors that influence both MAO and sensation seeking separately. However, Zuckerman's work remains intriguing. He also found that MAO gradually rises with age, which is certainly in keeping with the general observation that sensation-seeking and risk-taking behavior diminish as individuals age.

Of most interest to my research, however, are his observations about gender. Zuckerman found that the average woman exhibits less sensation-seeking behavior than the average man. Women's generally higher MAO levels support Zuckerman's observations. A word of caution here: these observations are about averages. There are many women who rank higher than many men on the sensation-seeking scale. As in most things, individuals vary considerably—no one would consider Joan of Arc, Amelia Earhart, or Harriet Tubman low-sensation seekers. However, on average, with everything else being equal, a woman will not rank as high as a man on the sensation-seeking scale.

A recent study at the Harvard Center for Risk Analysis takes a look at gender differences from another perspective. The researchers queried a random sample of people on their concerns about eight environmental hazards, including x-rays, ozone, pesticides, and global warming. Perceptions of hazards can influence decision making and "risky" behavior. The researchers found that hazard "believers," those indicating a high level of concern about these hazards, were largely women. Hazard "skeptics" tended to be men. On the whole, women consistently ranked the hazards higher than men did, a finding consistent with earlier studies.

The study then sums up the sociological and psychological explanations that can explain these findings. One focuses on the idea that women are more nurturing than men, leading them to be more concerned about hazards to their family. Another points out that women may find it easier than men to reveal their concerns, since men have been socialized to equate fear with weakness. Or perhaps women view science and technology with more alarm than men do because they are less familiar with it. In the study, researchers noted that the "believers" tended to have an egalitarian worldview, that is, a set of beliefs and values that are often critical of elitism. Such people might regard environmental hazards with great concern because the hazards disproportionately affect poorer members of society.

The researchers then returned to the data and took a second look, using a technique called multivariate analysis to factor out irregularities that could have skewed their findings. They discovered, for instance, that people who had children at home were more likely to have a heightened sense of concern about health hazards, to be "believers." And the researchers found that more women than men had children at home. Yet even when this factor and the "worldview" tendencies of women were taken into account, gender remained a significant factor in determining how people view health and environmental hazards. In fact, scientists don't yet know what factors — environmental, biological, or a mix of both — could be responsible for the gender gap regarding risk propensity.

In the broad evolutionary view, what selective processes historically supported risk taking? Risk taking on the surface might not seem to confer an evolutionary advantage. It's hard to see how risk taking advances the cause of passing along one's genes. A tendency toward risk taking might produce the opposite result. From the perspective of the community, however, risk taking takes on a new light. Sensation seeking (and its correlate, risk taking) could provide a critical advantage to hunter gatherers, for instance, by inciting them to fight, forage, and explore new territories. Exploiting new resources could mean the difference between starvation and survival for small bands of humans living close to the land.

By pushing off beyond the relatively safe and known confines, individuals could bring back items of enormous benefits to their community: new food sources, new mates to strengthen their genetic pool, new technologies from other peoples. Someone had to test the fruit or mushroom to see if it was edible or deadly poisonous, or to track down and kill a bear that was tormenting the community. Because the explorations involve venturing into unknown territory, the risks to individuals are high. It wouldn't make sense for everyone in the community to become involved in these activities. Psychologist Michael Apter suggests that the modern equivalent is the test pilot who must climb into the cockpit of an experimental jet fighter and fly it. Once the pilot has tested the basic performance and airworthiness of the plane, she may "push the envelope" to establish the extreme limits of its speed. Her report will help correct flaws and establish parameters of performance for future pilots.

But risk-taking individuals can also be destabilizing to a hunter-gatherer society. Their explorations or exploits could lead to unsettling, even revolutionary changes to the status quo. In contrast, the nonrisk takers act conservatively, being generally not interested in pressing out beyond the confines of their small, known area unless forced by necessity. Such behavior promotes stability in the short term. However, over the long term, this strategy could result in depletion of food sources and make a community vulnerable in the event of a natural catastrophe. Logic dictates that successful groups, those that survive and flourish, would carry a balance of both types of individuals. The tension between these two types would help keep a societal group or community in a healthy balance. As social orders grew more sophisticated, the risk takers would take on the roles of warriors and explorers, while the risk averse would become the healers, religious leaders, and storytellers. Many different job and responsibility levels could theoretically accommodate the range of risk tolerances found in a community.

The bottom line is that individuals carry vastly different responses to risk in their world. These are not just a matter of preferences and tastes but also a deeply rooted and intrinsic element of our character, personality, biological makeup, and societal organization. “For High Sensation Seekers,” says Zuckerman, “the world is interesting, full of opportunities for exciting new experiences.” To the low-sensation seekers the world is a dangerous place: risks are best minimized. “Low Sensation Seekers are not necessarily fearful people,” says Zuckerman, “they just don’t see the point in taking risks.”

Knowing where you fit on the risk-averse, risk-prone gradient is a critical element of learning the modern language of risk. When faced with the myriad risk choices that modern life presents, our individual predisposition toward risk has great bearing on how we weigh the outcomes of these decisions. Such knowledge may also reveal that differences of opinion regarding risk may not stem from the facts or concrete details but rather from basic differences in our biological and psychological makeup. As they continue their research, geneticists, biochemists, and psychologists are beginning to ferret out just how much genetic built-in responses determine our reactions to risk.

CHAPTER SEVEN
WAR ON THE SOMA

On a mountain-climbing expedition in Siberia, a support brace I stuck inside my plastic ice-climbing boot chafed an open sore into my shin the diameter of a soda can. Rereading a journal entry from the time, I noticed that I gave the injury short shrift, noting it more as a bother than anything else. In the context of climbing at high altitude deep inside Russia, the wound seemed insignificant. It posed little risk compared to other issues, such as climbing safely, not stepping on bad ice, or avoiding a fall. I didn't realize it at the time, but my body was teaching me a profound lesson in risk management.

The remote Altai Mountains stick up like knives along Russia's border with China and Mongolia. Soviet army helicopters had dropped us off at a high camp in a cirque for a couple of weeks. The living was spartan, the food lousy, the weather cold. I took it for granted that the wound would heal—my body had always taken care of such things in the past. But it didn't. It didn't bleed anymore, but neither did it dry up and scab over; while it didn't seem infected, it remained open. Back at home where I would be warm, rested, and well fed, it would have healed quickly. But up at 12,000 feet, where I had a bad cold, was getting a lot of strenuous exercise (and enjoying the awful Russian cuisine), my body had settled on a basic risk response — it had more important things to attend to than fixing the wound. With its limited resources, which were stressed at the time, my body directed its energies primarily to keeping me warm and alert. My body ordered its priorities, recognizing that its survival depended more on staying alert than on cleaning up a wound.

The human body has a remarkable ability to withstand significant threats from extremes of hot and cold, deep wounds, starvation, and stress. It does so by the skillful juggling of limited resources. If the skin is cut, a clotting mechanism engages to stop the bleeding, while the immune system cranks up to neutralize microscopic invaders that might enter through the wound. Blood brought to the area by inflammation whisks away debris as the injury heals. While these resources tend to the healing process, the body must also keep its eye on the big prize of keeping the organism alive by supplying oxygen through the blood to all cells, maintaining a balance of carbon dioxide and water, and keeping within a narrow temperature span. Meanwhile, organs and tissues work in concert to prevent injuries and toxic substances from causing damage. The liver and kidneys detoxify the blood. The intestines absorb nutrients. The colon expels dangerous materials. A network of tiny blood vessels known as the blood-brain barrier protects the brain from potentially harmful chemicals while allowing oxygen, glucose, and water to enter. When the body's defenses can't fight a virus, cancer, or severe injury and still keep the basic functions operational, it dies.

From the perspective of the body, life is a minefield, full of risks. The body is vulnerable even while reading this book. Say a person's leg falls asleep, cutting off blood supply. Or a bee flies through the window and stings his forearm. A person could rise too quickly from a chair, faint, and hit his head on the coffee table. He could inhale a virus lingering in the air from a sick friend's visit. In all these cases, the body must quickly and effectively gather its available resources and respond. To most of these mundane assaults, the body has evolved effective defenses. Considering all that could happen to it at any one time, the body's resources are like a small army defending hundreds of miles of open border with many hostile neighbors. Evenly spaced out, the troops would be so sparsely deployed as to be of virtually no use should an attack come. Yet if the troops are all massed in only one spot, other areas of the border could become vulnerable. The body deals with this dilemma by operating an efficient communication system and a rapid response team. Like a modern military organization that can quickly ferry troops in helicopters from one hot spot to another, this "army" makes its limited resources count.

In this light, my body's response to the altitude and wound made perfect sense. My story in Siberia ended on a happy note. While seated at the opening of my tent, a tough Russian mountain guide named Slava walked up. He had brown rotting teeth, indulged a penchant for vodka in the early morning, and exuded rock-hard self-reliance. Motioning for me to pull up my pant leg, he looked at the wound, then pulled a small vial from his parka. Out of the vial came a tarry, black material that he carefully daubed onto my leg. Within a day, the gunk had helped form a scab, giving my body's healing powers a jumpstart, and the wound dried up. Only later did I remember to ask him what was in that vial. "You know the bottom of caves?" he asked. I nodded, growing more alarmed by the moment as it dawned on me what he was about to say. As we worked through his broken English, I learned that the black material consisted of bat droppings, thousands of years old, scraped from a nearby cave floor.

The healer who applied the bat guano to my wound operates

under the same premise as does my family medical practicioner, although the latter has far more sophisticated tools at his disposal: both healers use treatments to bolster the defenses of an injured body. The emphasis on fixing the body supports a notion of the human body long held in the Western world. According to this view, the body is like a gallant eighteenth-century sailing ship that bears the soul on its voyage through life. The ship is generally seaworthy but sometimes in need of caulking and repair. Eventually, the ship grows creaky and old, leaks develop that are difficult to repair, the vessel becomes slower and less seaworthy, its sails fray, and the crew grows tired.

New evidence in the field of molecular biology makes the ship metaphor obsolete. While geneticists are working out the biological predisposition in favor of or against risk, molecular biologists are opening up a new frontier of risk in biology, not on the level of the organism, but on the level of the cell. Risk, when viewed on the cellular level, offers dramatic new insights into how our body responds to risk and promises to overturn old notions of how we evaluate risks facing our bodies.

Over the past couple of decades, bits and pieces of new information, often technical in nature, have trickled down from laboratories, scholarly journals, and scientific symposia into general circulation, and have begun to challenge this traditional view of the body, with major ramifications for how we understand and evaluate risk. To a large extent, individuals define physical risk according to how the human body responds to a threat or hazard. An activity, product, or substance carries greater risk if it can kill as opposed to cause an injury. Obviously, a game of soccer carries less risk than skydiving, not because kicking a ball around cannot result in injury — it often does — but because it's extraordinarily rare for soccer to cause death. The body has certain tolerances for surviving impacts, fighting bacterial infections, and other threats to it that are quite different from the tolerances of other life forms. In this regard, the human body is an important baseline for evaluating physical risk. Research on the level of the cell has begun to shift that baseline, forcing us to reexamine our overall understanding of physical risk. Familiarity with this shift is essential in speaking the modern language of risk.

For the first time in history, biologists can gaze right into the inner workings of body's smallest unit, the cell. What they've discovered is truly remarkable. For the first time, biologists comprehend not just how the whole body responds to risk but also how each of our 30 trillion cells reacts. Our cells, not just our bodies, are remarkable risk managers. Detailed information about how cells handle risk has altered thinking about how individuals should make decisions about health and evaluate true physical risk. Few realize how profoundly this quiet revolution has begun to change our general understanding not only of the body but also of risk, cancer, aging, and longevity.

This information reveals that life as a cell — second by second — is no picnic. Cells live in a state of siege. Bombarded from all sides and from within, they survive only by maintaining a high level of resistance. Life at the cell level is not a picture of health punctuated by bouts of illness but a ceaselessly violent world governed by the struggle for survival. This new picture supports a model of the human body that resembles an intergalactic space cruiser of the future rather than the sailing ship of yore. It's in a constant state of attack from the outside — meteors, cosmic rays — and inside from a nearly constant parade of hostile aliens who have escaped from the brig. Fortunately, the ship's shields are good, and the security forces are effective fighters.

The door to this incredible world cracked open in 1953 when biologists James Watson and Francis Crick published their seminal paper describing the shape of deoxyribonucleic acid, or DNA, the basic blueprint of the cell. In an organism inside each cell lies all the information necessary for life, all packed magnificently inside a space no bigger than 1/2500th of an inch. DNA's smallest units are base pairs, many of which together form a gene, the unit that can build proteins, such as enzymes, that control the myriad biochemical actions in the cell.

In their watershed discovery, Watson and Crick found that DNA took the shape of a double helix, two identical strings of nucleic acids joined together to form what looks like a twisting ladder. This seemingly simple observation stimulated a host of possible new insights into how the machinery of the cell operates. "It has not escaped our attention," they wrote, "that the specific pairing we have postulated immediately suggests a possible copying mechanism for the genetic material." Their prescient insights hinted at a pivotal point: DNA was like the big photocopying machine in the outer offices of a company president. From it come all the company directives that are distributed throughout the company. This observation would have tremendous repercussions for understanding life at the cellular level.

If DNA did in fact contain the information that a cell and organism needed, how was that information communicated from the DNA to the rest of the cell? During the 1960s, biologists worked out the details, identifying cell messengers, known as mRNA, that travel to various parts of the cell to tell the cell factories how business should be conducted. The analogy of DNA as a blueprint grew popular. I used to think of an architect, carrying around his plans coiled tightly under one arm. He walks the building site, sometimes referring to the plans as the building takes shape. While it does contain all the information upon which the cell and organism are based, DNA, according to new research, is far from a monolithic set of instructions carved in stone. In fact, the molecule itself is far from stationary and inviolable, but actually abuzz with constant change and movement.

Inside the nucleus of the cell, pieces of DNA regularly move about. Foreign molecules cling to it and change its shape. Crews of enzymes patrol the DNA, regularly clipping out pieces for repair. The amount of activity inside the cell is difficult to comprehend: imagine the streets of New York during a summer rush hour with all the fire alarms ringing. Researchers have discovered that the cell is under constant bombardment by all sorts of destructive agents. Each of the body's 30 trillion cells is literally fighting for its life every second. University of California at Berkeley molecular biologist Bruce Ames calculates that on average, each cell takes a "hit" from some destructive agent every ten seconds. That means that just in the time you take to read this chapter, the cells in your body will have been assaulted many trillions of times. These assaults often come unannounced, like a terrorist or guerrilla attack: quick rises in temperature, an influx of pollutants or heavy metals, a dose of radiation, or exposure to toxic waste products from run-of-the-mill cell metabolism. The cell is like the forts in those trite old westerns—always under a hail of arrows. At a hit every 10 seconds, an average cell receives 8,640 attacks a day, more than 3 million a year. And that's only one cell; remember, its neighbors are being attacked with the same ferocity. The science of molecular biology has also begun to reveal that the cell has an astounding arsenal of defensive weapons and tactics to fight this onslaught. In this microscopic world, legions of soldier enzymes repair cell damage; organized cell suicide occurs regularly; scavenger proteins prowl like hyenas on the African veldt.

Remarkably, the vast majority of the cell's enemies are natural, not the evil chemicals one imagines. One culprit is rays from the sun. Radiation, essentially energy in motion, exists throughout the natural world in the sun's rays, in rocks and earth, even in people's bones. The rays of the sun contain an invisible wavelength of ultraviolet light, against which we protect ourselves by applying sunblock or buying sunglasses with a protective UV coating. Every time natural light falls on the skin, even during cloudy weather, the body is deluged with ultraviolet light. Unless a person gets sunburned or feels woozy from sunstroke, the sun's effect on our bodies is not particularly noticeable. Yet it's quite noticeable on the cellular level. Ultraviolet light emits a so-called ionizing radiation, which means that it can strip electrons away at the subatomic level. This process interferes with the vital electric charges of atoms and can wreak havoc with the mechanisms of the cell. Such radiation is difficult to elude. Invisible and odorless radioactivity in the form of radon penetrates our basements. A person who spends a couple of hours in certain rooms of the U.S. Capitol and New York's Grand Central Station that were built with naturally radioactive granite will exceed the annual radiation exposure limits set by the congressionally mandated National Council on Radiation Protection and Measurements. Even sleeping with another person raises exposure, because all human bones contain a naturally radioactive form of potassium that emits faint traces of radiation into the environment. Other forms of ionizing radiation include microwaves and x-rays, to which our bodies are also exposed.

Even more devastating than radiation are products that the cell itself produces during its ordinary routine of cell metabolism. Just as energy production generates air pollution and factory processes produce toxic waste, the normal operations of the cell create toxic by-products. On the large scale, our organs filter, screen, and collect toxins created from the body's normal activities, such as isolating unwanted materials found in food. The harmful substances are efficiently eliminated through the urine and feces. This process is also going on at the cellular level. Each breath taken enriches our blood with oxygen, which then flows throughout the body and becomes available to each cell. Inside the cell, organelles that act as the center of energy production, the mitochondria, break down the oxygen molecules and capture the energy hidden within them. The final product is water.

Along the way, however, the process creates by-products such as hydrogen peroxide, superoxide, and hydroxyl radicals. Known as "oxidants" or "free radicals," these molecules have an unfortunate capacity to bond with other molecules in an often destructive process known as oxidation. Left unchecked, oxidants run riot inside a cell like the proverbial bull in a china shop. Molecular biology has taught us one of nature's most remarkable paradoxes: the very processes that sustain life also pose grave risks to it. And much of the miracle of the cell hinges on its abilities to fight these internal menaces.

Scientists once assumed that a dose of radiation or some form of toxin simply fried a cell upon contact and killed it. Modern molecular biology, however, reveals a far more sinister picture. While certain high levels of radiation certainly will kill a cell, much smaller and far more common everyday doses leave the cell largely intact, but can start something far more insidious than outright death. The radiation can disrupt the DNA sequence, critically damaging the encoded information that future cells need in order to function. Instead of killing one cell, it can ruin a whole chain, or many generations worth, of cells. It's akin to agents planting false intelligence information with the enemy: the repercussions can be devastating, not just to a group of soldiers but to the whole war effort.

Harvard University's Jac Nickoloff compares a cell's DNA to a movie. The scenes in the movie are represented by the approximately 100,000 genes in the genome. A scene in this incredible movie contains thousands of frames, each represented by a matched pair of four chemical bases. Each cell has two copies of this film, coiled together. Damage to the DNA can scramble a frame or cut out a few, altering the flow of the film. For instance, the chemical benzopyrene, found in cigarette smoke, is a large molecule that attaches itself to one of the frames of DNA (known as a base) and changes it to a different frame. Fortunately, there's another copy—the other side of DNA's ladderlike shape—so that repair enzymes can come in and splice in a corrected frame. Sometimes, however, these corrections contain errors, or the repair enzymes don't locate the problem. If both copies contain problems and frustrate the cell's ability to repair the error, then there's a risk of passing on an altered version of the film when the cell replicates. It's like a careless operator who copies only part of a film or jumps around while copying. The people who receive the duplicated film may not be able to make sense of it.

While a garbled film may be confusing, the impact of jumbled DNA can be disastrous for the cell and perhaps the entire organism. Should the mixed-up DNA disrupt the mechanisms that monitor and control cell growth and replication, cancer can follow. Every tissue in the human body is vulnerable to at least one form of the 100 documented types of cancer, all of which share the common trait of chaotic cell replication. Naturally, cells contain powerful weapons to stop or kill cells that show signs of out-of-control replication. It's in the cell's best interest, because chaotic cell growth can lead to malignant tumors. Should these unruly cells metastasize, that is, leave the tumor site and colonize elsewhere in the body, the cancer can cause death to the organism. The cell is so good at repairing itself that mutations in a single cell may take decades to bloom into full-blown cancer. In a 1995 article in *The Lancet*, a leading British medical journal, molecular biologists Dennis Carson and Augusto Lois wrote that "mutations, deletions, translocations, and duplications of more than a hundred different genes may contribute to tumor formation, depending on the tissue of origin." Half a million Americans die of cancer every year.

Most human cancers occur when two major classes of genes — representing only a small fraction of the entire genome — are damaged. The first type are proto-oncogenes, genes that direct the process in which a cell grows and divides to form new cells. Insults to these genes by radiation, oxidation, or pollutants can alter this process. The cell can spin out of control and replicate like mad, perhaps from manufacturing too much growth factor. The second type of gene is the tumor suppressor. These genes function like the emergency medical teams sent out to the scene of an accident. When something goes wrong, such as when a cell begins out-of-control replication, these genes turn on and pull the plug on cellular activities. Tumor suppressor genes can stop the cell from replicating, thus breaking a potentially deadly chain of bad cells. These genes have another especially effective trick, known as apoptosis. Should a cell show signs of mutations that could be detrimental to the organism, the cell will sentence itself to death and commit suicide. Little is yet known about how cell suicide is triggered, but the process is well documented. First, the cell removes itself from other cells. It then appears to boil. Blisters and bubbles appear on its surface. Then it implodes, its DNA and nuclear material condensing and breaking apart. The cell fragments are devoured by other cells.

Thus, the viability of an organism hinges not just on manufacturing new cells but also on shedding potentially dangerous old ones. Millions of your cells are killing themselves right now. The process is similar to a baseball team's manager ordering a sacrifice fly from a player to score a teammate: while the batter doesn't help his own batting average, his effort helps the team win. The success of chemotherapy and radiotherapy in battling cancer comes not from blasting the cancerous cells to death, but actually from inducing them to undergo apoptosis. The barrage of drugs serves to jump-start the body's own mechanism for self-preservation.

One of the hottest corners of molecular biology revolves around a single tumor suppressor gene, known as p53, but more impressively dubbed "guardian of the genome." Discovered in 1979, p53 at first attracted little attention, joining a long list of genes that appeared to cause cancer. But in 1989, researchers showed that it could suppress or kill tumors. In the subsequent rush of research on p53, researchers found that a damaged and inoperable p53 gene lies at the heart of much human cancer, including some 80 percent of colon cancer, 50 percent of lung cancer, and 40 percent of breast cancer. The p53 gene appears to be a major player in controlling cell suicide and cell division, some kind of genetic switch that flips on when the cell gets in trouble. In the rare cases where the healthy functioning of the body depends on the fast proliferation of the cells, the body has evolved mechanisms to prevent p53 from interfering. This happens when a pregnant woman's breasts begin to swell in anticipation of feeding a newborn and breast cells are proliferating furiously to get ready to produce milk. The p53 would normally red-flag this fast cell division as an error and initiate apoptosis. In this particular situation, the p53 gene is sent to the Siberia of the cell—the cytoplasm—so it can't interfere. Thus, the milk is produced for the infant.

Problems appear when a mutation occurs to the p53 gene itself; the benzopyrene in cigarette smoke, for instance, attacks a particular "hot zone" in the gene. On most occasions roving repair enzymes will correct the damage. If they don't, however, the next generation of cells could contain defective DNA. With the p53 gene not working, the cell may not be able to undergo effective damage control should things go wrong. This leaves the cell—and the whole organism—open to one of the most damaging human conditions: cancer. Douglas Brash of the Yale University School of Medicine explains this in the context of something most of us have experienced: sunburn. When sunburn occurs, dangerous ultraviolet light has bombarded our skin cells, playing havoc with the DNA. The p53 gene sounds a fire alarm, activating a mass suicide of the genetically impaired cells. The skin turns from bright red to white with scaly dead skin. New cells take the place of the old cells and, if the afflicted person has any sense, he or she won't expose them to ultraviolet light again in a hurry. When repeated exposure occurs, the chances rise markedly that the p53 gene in one of the skin cells will undergo dangerous mutations that are not corrected by the over-worked cell repair mechanisms. If the crippled p53 gene is copied into new skin cells, it may not activate when the skin is burned again. The cells, with the altered DNA, do not die and slough off, but replicate and pass on their bad DNA. This, Brash suggests, eventually causes the raised tan or reddish splotches on the skin that often occur on the face, scalp, and hands of sun worshippers. These patches can eventually turn into squamous cell cancer, a slow-growing skin malignancy that scientists at the American Academy of Dermatology estimate kills 1,500 Americans a year.

Alterations in the p53 gene can lead to more than skin cancer. For instance, studies of uranium miners exposed to the invisible but carcinogenic gas radon show that it often singles out a specific spot in the p53 gene for mutations. A potent carcinogen found in moldy corn and peanuts, aflatoxin B1, also appears to attack a specific sequence in the gene. Researchers at the National Institutes of Health have revealed that the position on the p53 gene where damage occurs could be an indicator of the source of the carcinogenic exposure. While damage to the p53 gene at certain points cannot now be used to identify a single carcinogen, information about where the gene has been hit could be a source of important clues about what caused it.

Armed with this knowledge about p53, researchers at Johns Hopkins University recently performed a piece of historical sleuthing. It's known that pieces of trash DNA, thrown out in the process of cell repair, exit the body through the urine. Sophisticated lab techniques now make it possible to find and isolate these tiny pieces of trash DNA from the urine. The researchers discovered a urine sample taken from then Vice President Hubert H. Humphrey in 1967, nine years before he died of bladder cancer. By using special gene-replicating technology, the Hopkins team found that Humphrey's DNA did contain a defective copy of the p53 gene. Intriguing possibilities arise from this finding. Had this technology been available in the late 1960s, Humphrey's cancer might have been caught early enough to prevent its spread. Lead investigator David Sidransky believes that a quick and accurate urine test for bladder cancer may be only a few years away.

Researchers refer to these recognizable alterations in molecular structure as biomarkers. In essence, a mangled p53 and other damaged genes or proteins can serve as tiny red flags, signaling that normal processes have gone awry. Now that genetic engineering techniques can locate these often tiny alterations—down to one alteration in a million DNA base pairs—biomarkers could serve as critical indicators of how and where cancer begins to grow. A damaged piece of p53 could be like the dead body in a murder, supplying critical clues to solve the mystery. Advances in molecular biology and analytic lab techniques have spawned molecular epidemiology, a science devoted to exactly this question.

Classic epidemiology for decades has revealed connections between exposure and various diseases. Examining large populations of people using carefully designed health questionnaires, epidemiologists have definitively linked cigarette smoking and exposure to such substances as benzene to various forms of human cancers. Epidemiologists have even quantified this information quite precisely, determining for instance that 11 percent of heavy smokers will get lung cancer. Before Crick and Watson opened the door into molecular biology, however, neither epidemiologists nor any other scientists knew the mechanisms by which cigarette smoke or other carcinogens worked on a cellular level. The burgeoning science of molecular epidemiology, a blend of classic epidemiology and molecular biology, looks at the progression of disease at the molecular level, using biomarkers to describe the cellular processes that lead from exposure to disease. This new breed of epidemiologist hopes to be able to police the molecular environment, apprehending the carcinogenic villain just as it begins to do its dirty work. By understanding the mechanisms of cancer, molecular epidemiologists may bring profound changes in our lives, causing us to rewrite our own sense of personal risk. There's every reason to believe that in the future a check up will include an analysis of our own personal biomarkers. Should they indicate nascent cancer, gene therapy or other remedies could nip the condition in the bud long before it grows deadly.

In the genesis of cancer, the alteration to the p53 gene appears rather late in the process. Is it possible to look back earlier? The answer, while complicated, appears to be yes. Molecular epidemiologists have discovered that carcinogens interact with a cell by bonding to a molecule of DNA or a critical protein. Researchers call the resulting complexes adducts. If a carcinogen forms a lot of adducts in a person's tissue, the possibility that a cancer could develop becomes more likely. Consequently, the accumulation of adducts could be an early warning system for cancer. Complicating the matter is the fact that the formation of adducts and response to carcinogens are by no means uniform among people. Some people smoke heavily all their lives or follow a poor diet and never develop cancer, whereas others who have never touched a cigarette and conscientiously avoid unhealthy food develop life-threatening cancers. The variations in our gene pool allow for many different responses to carcinogens. Rarely does a person carry outright "bad" genes, such as the BRCA-1 gene that appears to predispose its carriers to breast cancer. Over time, evolution selects against such hidden time bombs; the carriers of a bad gene tend to die off, leaving few chances to pass on the gene to a new generation. Researchers believe that only about 5 percent of the fatal cancers in America come from these bad genes. Instead, susceptibility to carcinogens is a far more subtle and complex affair. For instance, a person with a light complexion is more vulnerable than a darker-skinned person to skin cancer. One biologist states that there's as much as a 50- to 100-fold difference in tolerance among different people to various carcinogens.

By monitoring the formation of adducts, scientists could one day draw up a personal scorecard, a finely tuned summation of the cancer risks that a single person will face over his or her lifetime. What would you do if your summary indicates that it's 90 percent likely that you'll develop colorectal cancer if you consumed an average American diet and lived past 65 years of age? A person who walks to work would have to be quite reckless not to take an umbrella if the weather forecast predicted a 90 percent chance of rain. But how might a person act upon this kind of near certainty when it relates to something far in the future? Should that individual eat less animal fat and more roughage? Undergo frequent colon monitoring? Opt for surgery and partial removal when early signs show cancer? Or simply decide that the inconveniences of preventive actions, such as restrictive diets, are not worth a couple of extra years? These are complex risk decisions, heavy with weighted judgment calls and dependent upon a sophisticated understanding of both the science of probability and how cancer operates inside the body.

Because this information is personal, not general, it produces a shift in our relationship to risk. Today, most health warnings remain only indirectly connected to our lives, like a bothersome mosquito that flies about the room but has not yet landed on our skin. That's because the risks outlined refer to the general population, to the average person, most probably a 150-pound white male in his thirties who lives somewhere in the Midwest. Almost all of us can think of hundreds of ways in which we deviate from this average. That knowledge distances us from the impact of general health warnings. To some extent, as molecular biology has shown us, individuals do vary remarkably in their ability to cope with carcinogens. Witness America's worst cause of cancer, smoking, which still kills only one out of four heavy smokers with lung cancer. And individuals may believe that they eat better, exercise more, have a heartier constitution, or handle stress better than most other people, so that general health warnings don't apply. For many reasons, it's easy to shrug off advice and make our own rules as far as our own health is concerned.

When an individual receives personalized health information in the future, these rationalizations won't work. The person may choose to disregard the information but will no longer be able to plead ignorance. Imagine that a new baby is issued a biological report card along with a birth certificate and social security number. Molecular biology could one day soon determine which individuals will fall into the percentage who will get lung cancer if they choose to smoke. If your genetic profile fits into those 11 percent and you still smoke, you're making a deliberate choice.

The societal implications are immense. Each of us will have personal information that applies to no one else. An individual alone will have to decide how to act on this knowledge. Many of us will feel we cannot but take responsibility for our actions regarding our health. No longer can individuals blame the cigarette companies or anyone else for their lung cancer, for instance, unless it's because they were lured into addiction as immature people. In the final analysis, this new ability to learn about our genes signals a profound shift in how we regard risk. Decisions are thrown directly into our laps. As the mosquito that merely buzzed around the room now alights on our skin and prepares to bite, it's going to be up to us to slap it dead before it does.

The view at the level of the cell reveals the basic mechanics of life as a constant fight, a pitched battle against attack from all sides, both inside and out. In fact, the very processes necessary for life — cell metabolism, for instance — create risk for the cell. The cell survives and thrives in an environment where threats and risk are ever present. This conception of risk at cell level suggests that we should adapt a new strategy for how we think about the health of our bodies. While it remains important to cut out malignant tumors with a surgeon's scalpel or bombard a bacterial infection with antibiotics, this can't be the only strategy for staying healthy. Effort must be expended on prevention and maintenance, not just on fixing what's broken. Each of us must aid our body in its ongoing struggle by such measures as avoiding sunburn and paying attention to diet. Cancer and chronic disease rarely emerge from single events, but rather from exposure over time that eventually overwhelms the cell's defenses. Basic day-to-day preventive strategies can help the body in its fight against sickness.

CHAPTER EIGHT

WHERE DO WE DRAW THE LINE?

The most unsettling meal I've ever had took place not long ago in a benign setting. You might think it was the time I ate yak testicles (chewy) at an Explorers Club banquet or the time I drank fresh deer blood (frothy) with some Siberian farmers. Actually, it occurred at a tony restaurant in the shadow of the University of California at Berkeley, where I had arranged a meeting with Bruce Ames, a world-famous molecular biologist there. I had read a paper he'd written suggesting that every vegetable and fruit contains small amounts of nasty natural chemicals. To find out more about this unsettling idea, I paid him a visit.

For the past decade, Ames has made a habit of turning convention on its head. Perhaps his most startling notion is that the amount of carcinogens in food from pesticides pales in comparison with what's already there in nature. When I heard this information, I responded with the same strategy toward toxins that humans have always used: identify and avoid. I figured that I just could strike the bad ones right off my shopping list. But which ones? I figured dinner with Ames was a good place to start. A short man with glasses and the contrariness of a born scientist, Ames didn't waste much time. When they brought out our salads, he commented, "Cabbage contains 49 different toxins." Cross that off, I thought, moving the purple slivers over to the side of my salad bowl.

The satisfying aroma of cooked chicken met my nostrils even before the waiter brought my dinner to the table. But the thought of pyrolysis took the edge off my hunger. The process of cooking—pyrolysis—alters food, often dramatically. The smells and flavors that make cooked food so delicious come from the breakdown of amino acids, sugars, and other compounds in the food as a result of the heating process. These products recombine and form into a large number of new compounds that exude the smells and flavors associated with dinnertime but are also carcinogenic. Ames has written that three carcinogenic nitropyrenes found in diesel fuel are also found in grilled chicken. Overall, the browned and blackened food eaten over the course of an average day is at least several hundred times more carcinogenic than what's inhaled while living in a place with severe air pollution. Charcoal-broiled steaks contain billions of atoms worth of benzopyrene, number eight in the top twenty hazardous substances listed by the federal government's Agency for Toxic Substances and Disease Registry.

I began to lose my appetite as I thought of all the times I've barbecued. Despite the nature of our conversation, Ames relished the food in front of us. I began to get a little suspicious—is there something the matter with this guy? I thought back to another paper in which Ames reported that more than half of the forty-seven natural pesticides isolated from plants that have so far been tested on animals have been shown to cause cancer. Those twenty-five natural pesticides occur in apples, bananas, basil, broccoli, Brussels sprouts, cabbage, cauliflower, celery, coffee, fennel, horseradish, lettuce, mustard, nutmeg, parsnips, orange juice, black pepper, potatoes, tarragon, and more. A fair number of these foods played a role in my dinner that evening. Ames didn't even pause between bites to tell me that he has calculated that an average person eats between 5,000 and 10,000 different kinds of natural pesticides and their breakdown products over a lifetime—approximately 10,000 times the amount of artificial pesticides we ingest.

When the main course was cleared, Ames announced that "there are more than a thousand chemicals in roast coffee." Only twenty-six have been tested for carcinogenicity. Nineteen caused cancer in rodents. In fact, Ames calculates that there are more carcinogens in a single cup of coffee than in the amount of potentially carcinogenic pesticide residues ingested by an average person in a year. Scientists still haven't tested the thousand other chemicals in roasted coffee. I didn't even bother with dessert. A recent report from the National Research Council of the National Academy of Sciences convened an expert committee to evaluate many of the ideas first suggested by Ames. Their report, *Carcinogens and Anticarcinogens in the Human Diet*, found some fault with Ames's work but generally vindicated him. "It is plausible," the committee acknowledged in the careful wording of scientists writing for the lay public, "that naturally occurring chemicals present in food pose a greater cancer risk than synthetic chemicals." The report echoes Ames's assertion that "Ninety-nine point nine percent of pesticides in the American diet are chemicals that plants produce naturally."

The idea that certain plants contain noxious and sometimes deadly concentrations of poison has, of course, been known for a long time, derived from bitter human experience. It didn't take many accidental deaths for hunter gatherers to realize the effects of eating the death angel mushroom (*Amanita phalloides*), which contains amanitin, a potent neurotoxin. A 3,500-year-old document, the Ebers papyrus, lists a number of natural poisons—including hemlock, the ancient Chinese arrow poison aconite, and opium — that were widespread and well known. What is new is the knowledge that every plant contains a formidable arsenal of toxins in its makeup. Plants have evolved numerous chemical processes and defenses over many millions of years in order to survive. Trees, bushes, flowers, and ferns make use of growth hormones — for example, auxins, gibberellins, cytokinins — as well as color and aroma chemicals such as anthocyanins and monoterpenes. As stationary objects unable to defend themselves by flight, plants have evolved intimidating mechanical (think about the sharp spines of cactus) and chemical defenses against predators. Paleobotanists believe that amber, the hardened tree resin that is used to make popular jewelry, originated from a battle that proto-redwood trees waged with insect pests. Chemical signals alerted the tree to an insect boring into its bark, stimulating copious amounts of resin discharge to wash the predator away. Many plants developed a more subtle yet no less effective line of defense by evolving toxins potentially fatal to pests.

Celery, parsnips, dill, cloves, limes, and figs contain substances known as furocoumarins that appear to act as natural antibiotics, repelling attack by fungus. Furocoumarins irritate human skin and may have been responsible for cases where supermarket produce clerks developed blistering, rashes, and itching. Chrysanthemums contain substances called pyrethrins that are used in commercially available insecticides. In essence, every vegetable is just a unique collection of chemicals that have evolved to protect its survival. Allyl isothiocyanate gives the bite to horseradish, and at least eight different chemicals, including propyl mercaptan, cause our eyes to tear when we cut an onion. Chemicals contained in edible mushrooms (hydrazine), spices (safrole), parsley (psoralen), and bread (ethyl carbamate) have all produced cancers in laboratory animals.

Ames's research leaves me with the uncomfortable idea that if I cut out all the foods that contain toxins from my diet, I would have nothing to eat and would die of starvation. While the body's cells are well equipped to handle toxins, the idea that the food necessary for survival also contains poisons that can be injurious to our health appears a paradox of monumental proportions. If dangerous toxins exist in substances such as fruits and vegetables, how does that affect our view of artificial chemical additives?

During the last decade and a half, new findings in quantitative toxicology and a growing sense of how toxins affect our bodies and health have overturned our traditional understandings of toxins and the risks they pose. Considering that toxins and poisons in general—toxic sludge, pesticides, and herbicides, for instance—have become real point sources for discussions of risk, these new insights force us to reconsider how we regard the risk landscape. Despite the new information from Ames and others, most members of the American public still cling generally to old-fashioned notions about poison, thinking about toxic substances in black and white, usually in terms of whether they can kill a person or not. This now outmoded and simple notion can interfere with the effective evaluation of risk as we wrestle with some of modern society's most contentious risk issues: how do we regulate the products and by-products of the postindustrial age? What constitute safe levels of exposure to chemicals in our food, air, and water? What's natural? These issues can't be addressed effectively until individuals understand what new insights in toxicology have wrought. That's why a basic knowledge of these changes is critical to becoming fluent in the modern language of risk.

The impetus for recent advances in toxicology over the past decade and a half was the emergence of advanced quantitative techniques that made it possible for science to detect incomprehensibly minute amounts of compounds. Forty years ago, scientists measured the presence of a substance in parts per million. In other words, if a scientist examined 1 million molecules, he could measure the presence of a single foreign molecule among them. In contrast, scientists today can detect the presence of a substance down to parts per quintillion (a one with 18 zeros attached). That means that a scientist could detect the presence of a *jigger of vermouth* mixed evenly into a martini *the size of the Great Lakes*! The astounding leap into measuring the infinitesimal is akin to those startling electron microscopy images that reveal dust bunnies under the bed to be full of fierce-looking dust mites. A couple of decades ago, we didn't know certain substances were present because scientists simply couldn't detect them.

A scientist during an informal interview gave me a dramatic illustration of what this means. In the middle of our conversation he asked me to hold my breath for a moment. As I did, he began listing the molecules of the world's most deadly toxins I held inside my lungs: dioxin, arsenic, mercury, benzene. Arsenic is widely distributed in the environment, naturally occurring in the Earth's crust and in foods. Mercury is in the air at 2.4 parts per trillion. Benzene naturally occurs in volcanoes and forest fires, as well in the tissues of animals and plants. These substances have always been present in nature but have never been observed by science until very recently.

The ability to detect minute quantities has opened new frontiers and created some troubling implications. With portable monitoring devices in the home, for instance, toxicologists have discovered what's referred to as indoor pollution, that is, small concentrations of industrial chemicals that lurk in our everyday environment. Inside our homes in addition to naturally occurring chemicals may be found benzene from cigarette smoke, perchloroethylene from recently dry-cleaned clothes, paradichlorobenzene from toilet disinfectants and deodorizers, chloroform from the vapors created by hot showers. That's not to mention molecules of the glue used to hold furniture together, synthetic carpet fiber, and microscopic traces of pesticides tracked in on the soles of our shoes. Several scientists believe that these materials are currently far worse to individual health than industrial pollutants are. "If truckloads of dust with the same concentration of toxic chemicals as is found in most carpets were deposited outside," according to environmental engineers Wayne Ott and John Roberts, "these locations would be considered hazardous-waste dumps."

In addition to these new quantitative detection techniques, toxicologists over the past several decades have begun to make significant distinctions between one toxin and another in terms of their impact on the human body. Perhaps the biggest insight has been the recognition that hazardous materials can act chronically as well as acutely. The concept that toxins could work dangerously over the course of decades rather than minutes, hours, or days is relatively new. Toxins in their chronic form can form cancers slowly, scientists have learned, taking years to destroy the body and cause death. In addition to carcinogenicity (the uncontrolled growth of certain tissues in the body and the root of cancer) toxicologists have identified other toxins as mutagens (substances that cause changes, or mutations, in genetic material). Another class of toxins, teratogens, induces malformations in a developing fetus.

Toxicologists found that cancers could occur in many different parts of the body, depending on whether the chemical entered the body by contact with the skin, inhalation into the lungs, or digestion. Certain forms of asbestos cause lung cancer, for example, while radium affects the bone, where it's stored. Carcinogens themselves vary in potency. Compared with aflatoxin (a mold found in peanuts and peanut butter) cigarette smoke is a mild carcinogen. Carcinogens may also work in concert. For instance, the combination of breathing asbestos and smoking stimulates a cancer rate far higher than is caused by each individually. Research suggests that in some cases a carcinogen acts only as a "promoter," needing another "initiator" to cause carcinogenesis. In other words, one substance might not be harmful, but in league with another it becomes lethal. Recently, mostly during the latter part of this century, scientific understanding of the mechanisms of how toxic material affects human health has exploded.

The largest obstacle preventing the public from embracing these insights is that toxins by their very nature push our unconscious risk buttons. Unlike an arrow or bullet, toxins and poison are largely unseen, usually mysterious, and often tasteless and odorless. The universal sign of poison, the skull and crossbones, evokes the pirate flag, the sign of a fast-approaching and deadly vessel hellbent on malevolent action. As shocking as fatal car accidents are, they are understandable and usually explicable: wet highway conditions, bad tires, speed, alcohol, poor judgment. But with chemical exposure, the mechanisms are silent and unobserved, often poorly understood even by scientists. The area of toxins is ripe for misunderstanding, speculation, hysteria, and conspiracy theories.

Take ancient Rome's infamous Emperor Nero, for instance. Among other excesses and outrages, he is popularly held responsible for murdering his younger half-brother Britannicus with poison. During a sumptuous feast, Britannicus was offered a cup of wine, steaming hot as the Romans liked it. Fearful of poisoning, the royal family all took the precaution of using food tasters. Britannicus's food taster and loyal servant tasted the wine and pronounced it safe to drink, but too hot. Servants cooled the drink with water from a pitcher. Britannicus waved off the slave and took a gulp, passing the goblet on to his friend Titus. Just as Titus raised the goblet to his lips, Britannicus began to gasp for breath. Titus threw the wine across the room, fearful that the young prince had tasted poisoned wine.

As servants carried Britannicus out of the room, writhing in pain, Nero nonchalantly commented that his half-brother suffered only from an epileptic seizure and would recover soon. The nobles continued with their meal. Britannicus died soon thereafter. Had he succumbed to epilepsy or poison? Historians of the day roundly accused Nero of poisoning the water pitcher and arranging for the wine to be served scalding hot. The poison, wrote Tacitus, "circulated so rapidly in his veins that it deprived him of speech and life." Another historian reported that Titus too became ill for many days, because a few drops of the wine had touched his lips. Historians today cannot be sure whether or not Nero arranged Britannicus's death. Today, samples of Britannicus's hair follicles, a blood analysis, and an assay of fatty tissue could reveal the presence of toxins definitively, as anyone who's read a forensic mystery novel recently knows. Without chemical proof, the poisoning remains conjecture, ripe for legend and conspiracy theory. Whether or not Nero committed the crime, however, the act fits comfortably with his legendary ruthlessness, avarice, and selfishness. The nature of poison pushes our risk buttons and lends itself to such intrigue—and misinterpretation. The same remains true with modern toxins, such as Agent Orange, Alar, and dioxin.

In Nero's day, identified poisons were few, the dosages unclear. In fact, the Greek physician in his court, Dioscorides, took the first comprehensive approach to poisons, classifying them as derived from plant, animal, or mineral. Along with descriptions and drawings, he explored antidotes to poisons. In doing so, Dioscorides took a major step toward establishing the science, not just the art, of poison, which would one day become the discipline of toxicology.

The father of modern toxicology was a Swiss physician and natural philosopher who lived more than five hundred years ago. Named Theophrastus Phillippus Aureolus Bombastus von Hohenheim, he was better known as Paracelsus. His contemporaries were Nicolaus Copernicus, Martin Luther, Leonardo da Vinci, and other humanistic thinkers of the Renaissance. Paracelsus was an unusual physician in his day, a time when medicine was more art than science. He taught his student physicians to use chemical medications and not to rely on the potions then in vogue. He designed the first comprehensive mercury chemotherapy for syphilis. Before he died from wounds sustained during a bar fight in Salzburg, he made a simple but profound observation that defines the study of toxicology to this day: "All substances are poisons; there is none which is not a poison. The right dose differentiates a poison and a remedy." Put another way, any substance is deadly if ingested in sufficient amount: either water (10 quarts), sugar (2.5 pounds), or table salt (7 ounces) will kill an average 150-pound man who consumes this much at one sitting. The caffeine in 100 cups of strong coffee is fatal to a human being if drunk at a sitting. One hundred tablets of aspirin at a sitting are fatal, a fifth of whiskey often so. And 10 to 20 pounds of spinach contain enough oxalic acid to kill a person.

Certainly, most individuals drink enough coffee or eat enough spinach over a lifetime to kill them many times over; the reason they don't die is that the body processes the toxins over time. Some substances, such as iodine, salt, and vitamin D, are necessary for life in small amounts yet deadly in larger quantities. Two hundred grams of salt can kill an adult; two tablespoons can kill a one-year-old. Yet *without* 200 milligrams of salt per day we would soon be dead. Vitamin D is an even more extreme example. Were it not exempted, vitamin D would have to be labeled poisonous under the Hazardous Substances Labeling Act. Should a person eat 10 milligrams for every kilogram of body weight, that is, should a woman weighing 50 kilos (110 pounds) ingest 5000 milligrams of vitamin D, she could die. That compares to a similar dosage of the pesticide parathion. Yet *without* the trace amounts of vitamin D that are added to milk and found in vitamin supplements, children can get the debilitating condition known as rickets, which causes bones to grow soft and deformed.

Paracelsus's ideas turned into something called the dose-response relationship, a graph that plots the amount of a substance against its impact. Generally, the impact goes up in proportion to the dose, but some substances, such as arsenic, are just as toxic in small amounts as they are in large quantities. To make matters more complicated, toxicologists have discovered that large amounts of some things are not toxic, whereas small amounts are. On the surface, this may seem an impossibility. But the way a toxic substance is delivered to the system makes a large difference to its impact. Consider a child who breaks open a thermometer and eats the mercury. The body flushes it out of the system. However, mercury ingested in small amounts in, say, fish tissue over a long period of time can build up in the body and cause damage to the nervous system. Before science made distinctions such as these, our idea of poison was simple and direct, black and white. A poisonous substance killed a person in short order.

An observant British physician generally gets the tip of the hat for making the first causal link between exposure to a substance and cancer. In late–eighteenth-century London, Percival Pott, an early proponent of public health, observed malignant lesions on the scrotums of many chimney sweeps. He ruled out the conventional diagnosis of venereal disease and dug deeper. Pott published his observations in *Chirurgical Observations* in 1775. He described the terrible conditions under which the sweeps worked: they were often burned and bruised, continually covered in a thick grime of soot. The overseers kept the boys malnourished to ensure that they would remain undersized and could continue to fit into chimneys. Pott also observed that the sweeps washed irregularly and that soot accumulated in the folds of their scrotums. The soot, he concluded, was the cause of their cancers. Despite his elegant pleas, seventy-five years passed before England passed laws against boys cleaning chimneys by climbing through them.

Soot affected only a small minority, but food can harm an entire population. Because food is the major source of substances introduced into the body, the regulation of food additives and chemicals became one of the earliest forums for discussing efforts to limit physical risk. By 1900 some states had passed laws against certain food additives. During the Industrial Revolution, a host of new chemicals were discovered that uncaring or unknowing food manufacturers used liberally to doctor food products. In the eighteenth century, food producers used copper sulfate to make pickles green, vermilion (mercuric sulfide) to color cheese, and copper and lead salts to make candy look appetizing. Ground acorns filled out coffee, brick dust was added to cocoa. Popular outcry against these practices, heightened by shocking revelations about the meat-packing industry, ushered in changes in federal laws in the early 1900s. The rapid and bumpy changeover to an industrialized, primarily urban society helped spawn the Progressive movement, composed of reformers intent on correcting social inequities. Harvey Wiley, a leader in the arena of food safety, led a campaign for uniform federal laws aimed at ensuring food safety, targeting the use of boric acid, salicylic acid, and formaldehyde as food preservatives. His work led to the landmark federal Food and Drug Act of 1906 and the Meat Inspection Act. Wiley's crusade took many twists and turns, none more notable than the feeding experiments that the press referred to as Wiley's "poison squad."

Eager to learn about the ill effects of certain preservatives on humans, Wiley signed up twelve young employees from the U.S. Department of Agriculture, where he was the chief chemist. In the basement of the chemistry building, he built a kitchen and dining room. The all-male volunteers were required to limit their food intake to what Wiley provided. In addition, they carried around containers to catch all their feces and urine, no mean feat for a band of young men. Wiley and his various groups of assistants diligently cooked, weighed, and analyzed for a period of five years. Wiley began with experiments to determine the normal metabolism of the men. Following that he gave them preservative-laced food. His tests lasted for five years and involved many different sets of men and various preservatives. "It appears . . . ," wrote Wiley, "that both boric acid and borax, when continuously administered in small doses for a long period, or when given in large quantities for a short period, create disturbances of appetite, of digestion, and of health."

The press had a field day with Wiley's work, and the famous experiments even inspired popular songs. More than anything else,

Wiley's "poison squad" brought national attention to the issue of food additives and provided the impetus for the formation of the first federal food safety laws. Wiley himself came to believe that preservatives should not be banned, but should be used only when absolutely necessary. With a vision way ahead of his time, Wiley also believed that products should be labeled with lists of their contents and that food producers should bear the burden of proving that preservatives were safe.

Despite the major contributions that Wiley made to food safety, his experiments illustrate a major difficulty in the field of toxicology. By using human test subjects, Wiley's work was clearly unethical. He certainly didn't go as far as Catherine de Medici, who is described as the first experimental toxicologist in Casarett and Doull's authoritative textbook, *Toxicology*. The sixteenth century queen of France routinely prepared toxic concoctions and mixed them into food. In almost unimaginable acts of cold-bloodedness, she would then deliver her evil alms to the poor and sick. Carefully, she noted the effects on her unwitting subjects: how fast the toxins worked, what the victims complained of, how they died. How many men, women, and children she killed is not recorded. While Wiley certainly can't be charged with such flagrant disregard for human life, the fact remains that testing toxins on human subjects is considered to be unethical. This ethical proscription against testing on human subjects bedevils toxicology to this day. All determinations about toxins and human health can come only from circumstantial evidence — not an ideal scenario for scientists, who demand precise quantifiability.

When it came time for legislators to guarantee safe and sanitary food for the American public, these problems in toxicology bubbled to the surface. During the 1950s, a Senate committee struggled to determine permissible levels of pesticide residue in processed foods. This inquiry was breaking new ground in risk management. Charged with preventing health problems caused by chemical additives to food, the committee looked at the "endpoint" of cancer, though in the 1950s, science knew next to nothing about the biological and chemical mechanics of the disease. At that time, cancer appeared to be the most sensitive and subtle danger of toxins, and it often took years to emerge. The committee acted conservatively, not an unnatural or unwise strategy. Back then, FDA officials knew of only a few chemical carcinogens, which rarely occurred in food. Why not knock them out of the running, banning them from foodstuffs altogether? The committee's deliberations resulted in the Delaney Clause, which also found its way into laws on color additives in 1960 and on feed additives in 1968. This infamous amendment laid the groundwork for some of the damaging popular myths about risk analysis that continue to exist today.

Despite having no direct evidence of the link between pesticides and cancer in humans, the committee adopted a set of assumptions based on other evidence. First, they accepted rodents as reliable stand-ins for humans. If rats and mice got tumors after receiving doses of pesticide, they concluded, then these substances were a danger to humans, too. They set their threshold as low as possible: should any laboratory animal develop a malignant tumor from a substance, regardless of the amount administered, then that substance was not to be permitted as a food additive. This is known as the "zero risk" model, because no degree of risk is tolerated. The idea was to reduce the chances of cancer caused by food additives to zero.

When the Delaney Committee adopted its zero-risk policy in the 1950s, underlying it was a set assumptions known as the "linear, no-threshold" model. Any amount of carcinogen, no matter how small, will have a negative impact, equal to that amount. "No threshold" means that there's no point at which the carcinogen ceases to have a negative effect. Theoretically, that means that even down to a single molecule, a carcinogen will have a negative impact and potentially cause cancer. Because such innocuous substances, such as sugar, would flunk the test, a number of ingredients were grandfathered in as GRAS (generally regarded as safe).

The linear, no-threshold model appeared solid back in the 1950s, when scientists could measure substances only down to parts per million. Quantitative techniques acted as a sort of screen: if you couldn't detect it, it wasn't a problem. Since then, however, quantitative techniques have become so much better at tracing minute amounts of toxins that they have undercut the whole concept of zero risk.

If the linear, no-threshold model were strictly true for all these carcinogens, it's hard to imagine how humans could ever have survived as a species, much less thrived, over the past several hundred thousand years. As discussed in the previous chapter, the human cell constantly contends with hits from heavy metals, pollutants, radiation, and the by-products of cell metabolism. This knowledge, teamed with the revelations by quantitative toxicologists of trace amounts of the world's worst toxins throughout nature, suggests that at least some toxins have thresholds. Matthew Wald of the *New York Times* used the analogy of a river crossing to describe the dangers of assuming that doses can be extrapolated down to zero in a linear and even way. Assume that a river raging at 10 feet deep would claim the lives of 1,000 out of 10,000 people trying to cross it. Working the linear model down proportionately, you would logically expect 500 to drown when the river was at 5 feet deep, 100 people at 1 foot. Ten would die, extrapolating from this initial assumption, when the river was running only 1 inch. But such a scenario clearly defies reality. In some quantities, therefore, a chemical, like the river, won't cause any harm.

Unable to test human subjects, toxicologists struggled to fit this new information into theoretical models from the 1970s on. Fancy algorithms back up models with such names as probit, multistage, Weibull, multihit, and logit. Each model represents an attempt to grapple with the basic questions of dose and response, such as when, where, and if thresholds for toxins exist. Do all possible carcinogens have thresholds? If there are thresholds, at what point does the amount of a substance become significant, overwhelm the body's defenses, or set in motion the disruptive chain of events that lead to cancer or other maladies? There appear to be no easy answers.

None of these models can be verified, and depending on which one is chosen, a toxicologist can justify just about any course of action. The artificial sweetener sodium saccharin is a case in point. Analysis suggests that it's a mild carcinogen, because in high doses it causes tumors in rodents. Using the different toxicological models, a committee at the National Academy of Sciences projected the incidence of bladder cancer in humans who would use sodium saccharin between 1983 and 2053. Depending upon which model was chosen, the calculations ranged from the insignificant (0.22 fatalities) to the epidemic (1,144,000 fatalities). The inability to determine the nature of saccharin's carcinogenicity has resulted in a crazy ping-pong effect among legislators, who have changed their minds about its harmfulness repeatedly over the last two decades. Where do legislators draw the line? What's safe and what's not? How do we understand the effects of toxins in such small doses? Certainly, the Delaney Clause, with its zero-risk implications, has long outlived its usefulness. By nixing any additive that caused any sign of cancer in any given amount, the Delaney Clause lumped very harmful carcinogens together with ones that were questionable even in the huge doses administered to lab animals. In the case of sodium saccharin, the public has decided that the risk is worth taking for the benefits of enjoying diet drinks and food products.

Even when regulators and toxicologists agree that a particular substance is dangerous, often major problems exist in ascertaining the nature of the threat it poses to human health. No chemical has more confounded science and policymakers than the family of synthetic substances known as dioxins, which are incredibly toxic in minuscule doses. Toxicologists didn't even detect dioxins until the late 1950s, at the same time that Congress passed the Delaney Committee's recommendation for a zero-risk policy for food additives. Researchers discovered that dioxins were a by-product of manufactured herbicides, chemicals that kill or defoliate vegetation. The family's most toxic member, 2,3,7,8–TCDD, is 11,000 times more potent than sodium cyanide in killing guinea pigs. Animals exposed to only millionths and even billionths of a gram of dioxin often keel over, develop cancer, or develop abnormalities, making it the most potent carcinogen ever tested.

Searching for the existence of dioxin back in time, researchers dug cores into lake bottoms near industrial centers. They found evidence that dioxins had begun accumulating in appreciable amounts during the 1920s, which had peaked in the 1980s and had subsequently declined. The result of heavy industrial manufacturing, dioxin is a by-product of heating or burning materials containing chlorine. Incinerators, paper plants, and other industrial sources release these materials into the air and water. As more research was undertaken, scientists began to realize that dioxin was ubiquitous in the environment — in the air, water, soil, meat, dairy products, even mothers' fat-rich breast milk.

Of all toxins in the last two decades, none has created more fear and garnered more research dollars than dioxin. One top EPA official at the Office of Health and Environmental Assessment estimates that at least $500 million has been spent on dioxin research over the past two decades. The most notorious member of the dioxin family, known as TCDD, was isolated in 1968. On paper, the molecule looks, innocuously enough, like a short caterpillar. The body is made out of oxygen molecules and benzene; the two sets of antennae coming off each end are chlorine molecules. TCDD is a white powder that looks like salt but won't dissolve in water. In 1969, a study showed that TCDD caused birth defects in laboratory animals. Between 1962 and 1970, the U.S. military sprayed more than 11 million gallons of Agent Orange, a dioxin-containing herbicide, over Vietnam, killing vegetation along Viet Cong–controlled roads and pathways to deny the benefit of cover to enemy troops. Use of Agent Orange ended in 1970, but thousands of American troops had come into contact with the herbicide as aircraft, boats, trucks, and soldiers with portable canisters sprayed it into the air. Since 1979, VA hospitals have tested some 200,000 Vietnam veterans concerned about exposure to Agent Orange. Veterans have pinned a dizzying array of maladies on contact with the herbicide. Unfortunately, determining who was exposed and to how much dioxin is difficult to ascertain. Even so, epidemiological studies of Vietnam veterans have not found any evidence that exposure to Agent Orange caused or is linked to any of the various maladies suffered by veterans.

In 1976 a disaster in Meda, a small town in northern Italy, suddenly offered a potentially grisly open-air laboratory to investigate the effects of dioxin on human health, in a scenario that could have been taken right out of a science fiction novel. It began at lunchtime on an otherwise unremarkable Saturday in July, after production in the ICMESA factory, a subsidiary of the Swiss pharmaceutical giant Hoffmann–La Roche group, had stopped for the day and most workers had gone home for the weekend. The factory manufactured hexachlorophene, an agent in the production of insecticides, herbicides, and bactericides, in the heavily industrial area almost 20 miles north of Milan. That day, the temperature of a reactor rapidly rose, a safety disc dislodged, and a clearly visible chemical cloud escaped the plant. The prevailing wind blew the cloud southeast into the nearby town of Seveso, where several children noticed it and played in it. Before long, the cloud disappeared and everything looked the same again.

But it wasn't. In the intervening confusion of the next several weeks, birds dropped from the skies dead; chickens, rabbits, and dogs began to die; and people who had come into contact with the cloud developed the telltale sign of dioxin poisoning in humans: chloracne, a bad form of acne that strikes the face and arms. By then the scientific community realized in horror that the cloud had contained large amounts of the deadly TCDD, which was a by-product of the manufacturing process at the ICMESA factory. The residents were evacuated forthwith, leaving many of their household possessions. Police cordoned off 700 square acres with barbed wire. The enclosed area was divided into Zones A, B, and R, according to their varying amounts of exposure. In a controversial move, the Italian health minister authorized therapeutic abortions in the area because of the feared teratogenic effect of dioxin on fetuses.

Scientists held their collective breaths. No civilians had ever been exposed to such high levels of dioxin. All of the 724 people living in Zone A, the area of highest exposure, left their homes. In Zone B children and pregnant women were evacuated. On the surface, nothing appeared amiss. But the dioxin-laced cloud had settled on vegetation and soil, silent but present. Hysteria mounted in adjoining areas as residents who had not been evacuated grew concerned about their exposure. A fiery local and national political debate raged over what to do. Incineration of the topsoil at high temperatures would destroy the dioxin, but digging up the soil might just push the poison up into the air, permitting it to resettle. The church and other groups fought over whether allowing abortion to exposed pregnant women was ethical.

Overall, surprisingly little happened. There were no mass deaths or instances of deformed babies. Those with chloracne didn't develop cancer. The number of stillborns did not rise above normal rates. Almost two decades after the Seveso disaster, an Italian research team lead by Pier Alberto Bertazzi of the University of Milan examined the medical records of more than 37,000 men and women exposed to the dioxin cloud at Seveso and compared them with a control population of 180,000 Italians who lived nearby but had not come into contact with dioxin from the plant. He found that the 724 people living in Zone A showed no increase in cancer rates. Bertazzi noted, however, that this group was not a good source of information because of its small size and because of their brief exposure: they moved relatively quickly after the event. In Zone B, where nearly 5,000 people had experienced prolonged exposure to the dioxin, the researchers found that women suffered five times the normal rates of multiple myeloma (a rare bone marrow cancer) and gallbladder cancer. Men suffered twice the normal rate of liver cancers and more than five times the normal rate of certain cancers of the blood. Yet, surprisingly, the overall cancer rates for women were actually lower than normal and only slightly elevated for men. Curiously, breast and endometrial cancers for women in Zone B rang in at lower-than-normal rates. This evidence points up some other potential effects of dioxin on the reproductive system. Few scientists believe we can dismiss the impact of dioxin on human health as insignificant. Too much evidence hints to the contrary. Clearly, dioxins aren't your garden-variety carcinogens. However, the inability of the best toxicological, epidemiological, and chemical minds in science to determine the impact of dioxins on human health underscores the immense difficulty of evaluating the risks of carcinogens in general. Further muddying the waters are recent discoveries that dioxin appears to exist in uncontaminated deposits of ancient clay throughout the South. These findings, in addition to the fact that industrial emissions of dioxin don't account for all the dioxin found in nature, have prompted investigators at the EPA and some independent industrial ecologists to theorize that dioxin is also found in natural sources.

The Seveso incident points up something that has long worried toxicologists. In many cases, animals react differently from humans to chemical substances. The doses of dioxin they received killed some animals in their tracks, whereas in humans, the only definitively proven impact of dioxin was the severe chloracne rashes, although chronic impacts may yet be proved. Until recently, animal testing remained the single standard for determining whether a substance was carcinogenic.

In some ways, rodents are excellent stand-ins for humans because substances known to cause cancer in humans—such as aflatoxin, asbestos, benzene, and radon—also cause cancer in these animals. But simply extrapolating from animals to humans creates major problems. For one thing, test rodents live in a uniform, controlled environment where food, temperature, light, and stress levels are all standardized. In contrast, humans rarely conform to averages, eat wildly different foods, and enjoy vastly different lifestyles. The metabolisms of rodents are obviously different from those of humans. Cyanide doesn't affect rats but is deadly poison to humans. Tragically, the sedative thalidomide had little impact on rodents but created serious disfigurement in human fetuses.

Perhaps the most difficult problem with animal testing lies in sheer numbers. If a substance causes one fatal cancer in 1,000 animals (a high rate, most experts agree), it won't necessarily show up if you test 100 animals. Administering doses of possible carcinogens, adjusted for rat body weight and fast metabolism to resemble normal human consumption, becomes tricky when many of the carcinogens under investigation exist in parts per million or even billion. Optimally, the tests would require tens of millions of animals, an impossibility. Instead, scientists compromise by giving the laboratory animals megadoses of potential carcinogens over their short lives. That's where scientists complain that, for instance, in the study of the artificial sweetener saccharin, rats were given the equivalent of tens of thousands of cans worth of diet soda, far more than any person could consume over several years. The major criticism voiced by scientists against the megadosing strategy is that these large doses can cause massive tissue damage and subsequent cell proliferation that in and of itself leads to cancer. The high frequency of positive results found when testing with megadosing may thus be inflated because of the amount of the toxic material involved rather than the nature of the toxin itself. Recall Paracelsus's conclusion several centuries ago: everything is toxic in some amount; it just depends on the dose.

These concerns about the validity of animal testing, along with recent inroads into understanding the mechanisms of cancer, have prompted the EPA recently to restructure the way in which it now calculates cancer risks. This radical move could have far-reaching consequences on regulatory policy over pollution control, industrial emissions, the use of pesticides, and standards for cleaning up toxic dumps. The EPA proposes dropping its reliance on animal tests and using risk assessment instead. Analysis of a pollutant or industrial chemical will now include molecular biological information, such as how a particular chemical might alter DNA or impact how a cell regulates its metabolism. The guidelines will include information about a substance's chemical and physical properties, make distinctions based on how the agent is introduced (i.e., inhaled, eaten), and acknowledge the differing susceptibilities of exposed populations, such as children. The burgeoning of information on the mechanisms of cancer has enabled scientists to focus not just on whether a chemical induces a tumor at a certain dose but also on just how an agent might cause that tumor.

The EPA's new guidelines acknowledge that carcinogens are not all the same strength, nor all similar in the way they act. Risk assessors are now encouraged to use nonlinear methods at small doses and look for thresholds. Along with these changes at the EPA, legislators recently killed the long-standing Delaney Clause. Other factors, not just biological science, now enter the risk equation. Should risk be calculated differently for different populations? While the regulators struggle to define risk anew, however, most Americans still labor under age-old notions of poison and want simple answers. Is a substance bad or good, toxic or not?

At the same time that the Delaney Clause was being phased out over the last several years, new research into the effect of artificial hormones in nature suggests that we may need to revise our estimation of the risk caused by artificial and natural chemicals. A number of studies, which are by no means conclusive, suggest that commercial chemicals found in pesticides, plastics, industrial solvents and wastes, and petroleum products may accumulate in biological systems and disrupt hormonal balances in humans and animals. Glands in humans and most animal bodies secrete such chemical substances as estrogen, androgen, and thyroid hormones into the blood stream to regulate an array of bodily functions, including reproduction, sexual development, and metabolism. These commercial chemicals, so-called endocrine disruptors, may mimic the body's natural hormones, altering normal functions and leading to low sperm counts, a high incidence of birth defects, and breast cancer. Whether chemicals in fact disrupt the endocrine system on a large scale remains one of the most contentious questions in environmental circles today.

The EPA and Congress are moving toward mandating widespread toxicological screening and testing of the many thousands of chemicals used in society. These include naturally occurring chemicals as well as artificial ones. For the first time, new quantitative technologies that are relatively inexpensive and fast make this testing possible. If chemicals are identified as potentially harmful, and solid research does in fact support the idea that endocrine disruptors are altering biological systems, then a slew of difficult economic, ethical, and environmental questions will arise. It will become ever more important to speak the language of risk, because many of the answers to these questions will delve right into the heart of what we value and don't want to lose.

The risk decisions we will have to make will be more difficult than environmentalists or industrialists want to admit. When I reminded Bruce Ames about the devastation that the insecticide DDT wrought on the seabirds of America's East Coast, he countered that DDT saved millions of lives in the Third World by preventing mosquitoes from spreading disease. "Pesticides," says Ames, "may be one of the greatest boons to our health of all time." The logic is simple: pesticides contribute appreciably to the availability of better-looking and less expensive fruits and vegetables. Consequently, people eat more of them and are correspondingly healthier. The trade-offs are not black and white. Will we be willing to forego some of our lifestyle advantages so as to avoid the use of certain chemicals? These are matters of risk that we will have to decide when the time comes.

Regardless of what research tells us in the future, individuals need to adopt new risk strategies when it comes to thinking about toxins. For one thing, the old "identify and avoid" strategy doesn't work as well anymore. Many of us today skim newspapers for reports of what drugs, foods, and activities cause cancer. I for one file these way, vowing to avoid those substances in the future. Anyone who's tried to keep track of what's cancer-causing in this fashion knows that the list quickly becomes large and unmanageable. There are things that you shouldn't eat when you're old, young, pregnant, a drinker, on antibiotics, or have high blood pressure. Sometimes I don't even feel like reading the newspapers, so certain am I that I'll see the findings of another study that I'll have to add to the database of dos and don'ts in my head. Revelations in the field of toxicology necessitate a different strategy. It's no longer a matter of avoidance, but rather of managing toxic chemicals, as well as recognizing that all substances are not created equally as far as toxicity goes. In other words, it's a matter of balancing risks.

CHAPTER NINE

THE MYTH OF ZERO RISK

As I brushed my teeth one morning, I glanced at the label on the tube of toothpaste. It said: "Warning: Keep out of the reach of children under 6 years of age. If you accidentally swallow more than used for brushing, seek professional assistance or contact a Poison Control Center immediately." I must admit that for a few moments the toothpaste in my mouth felt like a nasty cleaning solvent. I might expect a warning like this on something in my garage, not on something I put into my mouth twice a day.

The Food and Drug Administration (FDA), the governmental organization that polices the food supply, instituted the warning because most modern toothpastes contain fluoride. Aside from rare cases of allergic reactions, too much fluoride can cause temporary bouts of diarrhea and nausea. For decades, fluoride has been dropped into the public water supply and added to soft drinks and canned foods to help prevent tooth decay. The FDA's intent was to increase consumer awareness of the possible side effects — not a bad idea, although the language bordered on the extreme.

Of course, the warning pushed my risk hot buttons, although I had no choice but to use the artificial chemical if I wanted to brush my teeth that morning. But the warning label unsettled me for other reasons. Despite what I learned working on this book, I still find it difficult to accept the idea that steps taken to protect me against risk increase my exposure to other risks at the same time.

In the case of the fluoride, the trade-off is increased dental health for everyone, allergic responses for a handful. An essential, and perhaps the most critical, foundation of the modern language of risk is that whenever an action is taken to reduce a risk in one area, a risk or risks in another area will be heightened. Eating vegetables to maintain a healthy diet also means ingesting trace amounts of toxins. Putting up bars on windows to protect against thieves makes it more difficult to leave the house in case of fire. Washing the hands with soap, which may lessen a person's chance of contracting a cold or flu, can also put the individual in contact with cosmetic additives that might be carcinogenic. A person who keeps a handgun at home to protect against intruders significantly increases the chances that someone in his or her family will be shot accidentally. Or a city's aggressive new crime program, for instance, might reduce the funds available for a program that screens children for lead poisoning. Consequently, a reduction in crime may accompany an increase in the number of children reported with lead-related health problems. The one overriding lesson from the mountains of literature generated by the science of risk analysis is that a risk cannot be assessed and managed along only one dimension, like a mountain peak waiting to be conquered. In fact, risk management, both individually and collectively, involves a world of trade-offs and demands a certain finesse.

I like to think of this new relationship to risk in terms of a snake slithering in the woods outside a person's house. Not sure whether the snake is poisonous or not, the homeowner can eliminate the risk of a family member being bitten by killing it anyway. For most of human history, that's what people have done: attack risks head-on by going after the snake. That's the way we've ensured our survival, by identifying a hazard and neutralizing it; and so humans have moved from hazard to hazard. But now the science of risk analysis, along with advances in ecology, enables us to look at the risk that the snake presents in a new light. Biologists have realized that the natural world is an interconnected web of dependent relationships, many too subtle for us even to recognize. Killing the snake does eliminate the immediate threat of a snake bite, but suppose this snake also played a critical part in keeping the wood's rodent population in check. Without the presence of this predator, the rodent population could grow, harbor nasty pathogens, and expose the family to the risk of disease that represents a far greater health concern than the risk of snake bite. Alternatively, killing the snake might make way for new predators that are even more poisonous.

While killing the snake may still prove to be the right thing to do, the decision is more complicated when this new information and way of thinking about risk are factored into the equation. Battling risks appears to be a linear process, but often it is not. An attempt to control one risk heightens other risks. Minimizing one of these new risks raises yet others, in a chain reaction that has no end. Attempting to reduce risk is like squeezing a partially filled balloon: you can make it contract under your grip, but by doing so you make it expand in other directions. Unlike squeezing the balloon, however, controlling one risk doesn't result in a clear and equal response in another direction. This is why managing risk in modern times has become critical. In dealing with the snake, the homeowner, like all of us, has a chance not just to respond to risk but also to *manage* it.

Simple as this concept may seem, many people still do not grasp it. The remarkable victories of the twentieth century against threats such as smallpox have been so convincing that they've helped create an illusion that alleviating one risk doesn't involve trade-offs with others. The price of a few adverse reactions to smallpox inoculations doesn't even register when measured against the vast number of lives saved. These successes have built a conceit that if enough resources and persistence are brought to bear, society can use its tools of science and technology to conquer all risks, as though they are merely serial obstacles waiting to be overcome. I call this the myth of zero risk.

For much of human history, working linearly toward zero risk — killing the poisonous snake — proved a workable strategy. In modern society the risk landscape has shifted so radically that this approach to risk has become the largest roadblock to individuals trying to learn the new language of risk. Gone now are many risks that faced Americans at the beginning of the twentieth century. On a graph, the line representing risks to our mortality has plummeted during this century. Now it has leveled out. Nothing society can do, other than artificially engineering a body that will live beyond 120, will make the line drop as precipitously as it has in the past. The risks that are easiest to tackle — in the sense that the risks and benefits are clear — have been moderated successfully, leaving us to wrestle with smaller, harder-to-define risks. This is why every one of us must become a good risk manager.

Automobile safety provides a good example. As mentioned in Chapter 1, traffic safety engineer Leonard Evans at General Motors Research Labs has calculated that a person wearing a seat belt is 42 percent less likely to be killed than one who doesn't. The trade-offs in terms of cost and increased risk — a person's inability to get out of the vehicle quickly in a fire, for instance — don't come close to offsetting the tremendous gains in safety seat belts provide. Air bags prove a little more troublesome in this respect. When used in addition to seat belts, air bags reduce overall risk by 47 percent, thus improving the odds only marginally over use of a seat belt alone. The trade-offs — much greater price, problems with accidental deployment, and danger to children and short people — make the safety gains less clear. Additional safety measures, such as special bumpers and helmets, appear not to be worth the risk downside of extra cost and inconvenience. As safety engineers, legislators, and regulators move down the risk ladder searching for ways to decrease risks, they encounter risk-benefit situations that are all shades of gray, not black and white.

The science of epidemiology serves as another example. This century epidemiologists have identified many major health threats. Smoking raises the chance of contracting lung cancer by as much as 3,000 percent. The health risks of obesity, alcohol consumption, ionizing radiation, and exposure to certain forms of asbestos are also unambiguous. Epidemiologists agree that they are unlikely to find any more risk associations of this magnitude. When epidemiologists push into second- and third-tier health hazards, the risks become more difficult to pin down, and the associations become weaker and possibly suspect. According to a 1995 special news report in *Science* titled "Epidemiology Faces Its Limits," most epidemiologists would question any single study that reported a potential new cancer unless the suspected agent increased a person's risk by at least threefold.

Many recent such "discoveries" report risk ratios well under this amount: a high-cholesterol diet (1.65 times normal risk for men of contracting rectal cancer), consumption of yogurt at least once a month (2.0 risk for ovarian cancer), regular use of high-alcohol mouthwash (1.5 times the risk for contracting mouth cancer), consumption of red meat five or more times a week (2.5 times risk for colon cancer). While these activities may represent legitimate health risks, taking preventive steps to minimize these more subtle risks may raise other risks beyond our comfort level. Not drinking alcohol in moderation could reduce a person's ability to handle stress and put his or her heart at risk. Cutting out saccharin, a health risk that's proved difficult to determine, may cause some people to gain weight, putting them at increased risk for a whole series of potential maladies. No longer does the model of knocking risks down one by one make sense. Attacking risks now involves a close examination of the other risks raised by moderating the initial risk—in essence, a balancing act.

In addition, the fabric of society has grown far more complex and interconnected, making it impossible to think of managing risks singly. The twentieth century witnessed the introduction of new technologies with such far-reaching and potentially catastrophic capacities that the stakes have changed. The process of scientific and technological advance, in part fueled by the desire to combat risk, creates hazards that did not exist before. What occurs is a paradox of modern times: our very success in minimizing risk has itself created new orders of hazards that require our attention. And these new risks, which often take years or decades to understand, in turn require further risk-management strategies, resulting in a never-ending chain.

Consider, for instance, the implications thrust upon modern society by an accidental discovery made by the French physicist Henri Becquerel and a young graduate student named Marie Curie, a discovery that would alter forever the way Americans regard risk in their lives.

During the 1890s, while a physics professor at the Ecole Polytechnique, Becquerel became intrigued by the recent discovery of x-rays. He wondered whether luminescent minerals might not emit x-rays too, if first exposed to the sun. To test his idea he placed mineral samples on photographic plates wrapped in lightproof paper and then exposed them to the sun, intending to examine the developed plates to see whether the material was emitting any of the curious rays.

One day in February 1896, he began preparation in his laboratory for the next item on his list, uranium. The weather that day didn't cooperate, however, so he postponed his experiment for another day. When he got back to the lab, he decided to develop the photographic plates that contained the uranium even though they had not been exposed to the sun. He was astonished to find the outline of the uranium on the plate. Less than two weeks later, Becquerel published a paper describing a new form of radiation that didn't need the sun and could penetrate paper and certain thin sheets of metal.

Marie Curie, a bright graduate student at the Ecole de Physique et de Chimie Industrielle in Paris, read Becquerel's paper, decided it was worthy of further study, and made it the subject of her doctoral dissertation. She and her husband, Pierre, discovered that pitchblende, a brownish-black, uranium-bearing ore, emitted stronger radiation than uranium alone. She called the emissions "radioactivity." The couple soon found that the pitchblende contained a new element, which they dubbed radium. Unaware of the dangers, Marie Curie spent the next four years cramped in an unheated outdoor shed in a laborious attempt to isolate radium. From one ton of pitchblende procured from a mine in the Belgian Congo, Curie procured 100 milligrams of radium, an amount about the size of a pea. Her physical labors were herculean, the dangers in retrospect monstrous. "One of our joys," she wrote "was to go into our workroom at night; we then perceived on all sides the feebly luminous silhouettes of the bottles or capsules containing our products." She then compared the light to "faint, fairy lights." The notebooks she used are still highly radioactive. In 1932, Marie Curie died of aplastic anemia, a form of leukemia probably caused by her exposure to radiation and those peculiar fairy lights. A Pandora's box of the most serious kind had been opened.

Radium found some promising uses in medicine, including some success in treating tumors. During World War I, contractors put the radium to immediate use, painting thousands of airplane instrument dials and wristwatches with luminescent radium paint. After the war, radium paint decorated many consumer items, from dolls' eyes to gun sights. By the mid-1920s, however, many of the dial painters began to experience cancer of their teeth and jaws. Some of them died. During their work, the dial painters, who were usually women, licked their brushes to create a fine painting tip. Each time, they ingested trace amounts of radium. Radium paint was considered harmless by scientists and physicians because it contained only minuscule amounts of radium. The horrible deaths and injuries that mounted proved otherwise. As with any newly discovered hazard, questions of all kinds surfaced. How much radium was safe to ingest? How do you monitor the safety of people working with this material? Eventually, however, the evidence pointed convincingly to the grave dangers of human exposure to radium.

During World War II, scientists in the Manhattan Project, the government's top-secret program to develop the atomic bomb, studied the effects of radium and the newly discovered element plutonium on animals and extrapolated safe exposure levels for humans. Safety precautions ensured that workers were not overexposed to radiation during the project. But of course, even as the scientists took risk-management steps against radiation, they opened up another Pandora's box with the atomic bomb itself. At first, the scientists working on the project considered the bomb only as an immensely powerful version of a conventional explosive. They soon came to realize that unlike a traditional explosive, an atomic bomb also spread an invisible mist of radioactivity known as fallout. Long after the violent blast, the radioactivity would linger, contaminating a wide area.

Three decades after the bomb blasts at Hiroshima and Nagasaki, new theories emerged that massive nuclear bomb detonations could cause nuclear winter, that is, microscopic particles sent up into the air could block out the rays of the sun and cause most life to die. The knowledge dawned that humankind might hold the power to extinguish itself from this planet. Americans felt this tangibly during the 1983 showing of the ABC teledrama *The Day After*. Some viewers grew physically ill from watching the fictional dramatization of the devastation wrought by a nuclear blast upon an average American community. The government's policy of instructing children to hide underneath their school desks, which had long appeared absurd, now seemed like a cruel joke. In a large-scale nuclear incident, there is no place to hide. In terms of risk — that is, the possibility of loss or harm involved — the stakes had jumped several orders of magnitude. Risk in modern times is an animal that keeps changing shape and size.

In other arenas of modern society, technology also operates on a far more magnified scale: immense dams hold back millions of gallons of water; passenger jets speed through the skies at 600 miles per hour at 30,000 feet carrying 300 people; nuclear power plants split atoms in the process of fission; complex industrial plants produce hundreds of chemicals and their by-products. The possibility of catastrophe in these situations, while extremely low, carries immensely high stakes; in some cases potentially millions of people could be killed, injured, or sickened. Viruses once locked in deep jungles can now jump around the world in a matter of days through the network of air carriers. Pest species formerly confined to a continent or island can now travel as stowaways on board ships that ply the oceans with increasing speed and frequency. With advanced marketing techniques, swift mass-production capabilities, and rapid distribution, a food product or drug could be used by millions of people in the course of only several years. Witness aspartame, marketed as NutraSweet and approved in 1981 by the FDA. It's now found in thousands of products and used by hundreds of millions of people across the world. Technology has become far more multifaceted and difficult to monitor. Risk experts point out that 50,000 consumer products are available in the United States. All told, about 70,000 different substances are found in the workplace or are sent out into the environment as pesticides, fertilizers, or various forms of industrial effluent. The National Institutes of Environmental Health Sciences and the EPA believe that toxicological information is available on less than 20 percent of these. "If even a small fraction of these presented the legal and technical complexities engendered by saccharin or flammable sleepwear," writes Carnegie Mellon's Baruch Fischhoff and his colleagues, "it would take legions of analysts, lawyers, toxicologists, and regulators to handle the situation."

Humankind holds the power to set global forces in motion by burning fossil fuels and causing global warming, opening holes in the protective mantle of ozone in the atmosphere with aerosols, triggering massive extinctions and fundamentally altering biodiversity through aggressive land development, and snuffing out most life forms through the deployment of broad-ranged nuclear weapons. As technology has brought us new wonders and horrors both, modern society has struck a Faustian bargain. In the twentieth century, Americans have gained a better standard of living, longer life expectancy, and greater individual health. In exchange, we have shouldered a new order of low-probability, but high-impact risks that the world has never seen before. This is one of the reasons why a Marsh and McLennan survey found that 78 percent of Americans believe they face far more risk than their parents did — even though they are living far longer and healthier lives. These new orders of risk have wrought havoc with our traditional ways of understanding and thinking about risk, particularly the idea that we can eliminate it altogether.

Studies in risk analysis suggest that certain risks of modern society may never be minimized or eradicated no matter how much effort science exerts on their behalf. This has become evident in some surprising research into the origin of large-scale technological disasters, an infrequent but highly visible aspect of modern times. Incidents such as the disaster at the Union Carbide chemical plant in Bhopal in 1984, the explosion of the space shuttle *Challenger* in 1986, or the oil spill from the *Exxon Valdez* tanker in 1989 are the most prominent examples. More frequent are airliner crashes. Every time a passenger jet crashes with major loss of human life, the public follows latest developments in the investigation by the National Transportation Safety Board. Journalists speculate about the contents of the "black boxes" and whether wind shear, pilot error, or terrorists are to blame. As experts and technicians explore various explanations, the public hangs on every new development. In the rubble and wreckage of a disaster, each minute piece is examined as if it were a talisman. I often fantasize that if the media and the public gave equal attention and the same intensity of focus to child abuse or highway accidents, we could make great and rapid progress in solving these problems.

The search for why things go wrong offers us a semblance of control over these horrible events. Hollywood thrillers have added to the notion that at the root of things gone wrong are terrorists, mad scientists, and industry captains motivated by greed. The searching process reassures us that lessons can be learned. Surely in the future, society can take steps to make sure a similar tragedy won't occur again. If a terrorist bomb caused the crash, better security in airports could stop a future terrorist from planting an explosive device. If it was pilot error, perhaps better training programs could be instituted. Or if quirky weather patterns are to blame, air traffic controllers and pilots could learn to recognize wind shear conditions. Perhaps an accident might reveal a design deficiency that can be remedied.

This search for answers is predicated on the idea that there must be a course of action that could have been taken that would have prevented the crash. Find it, and appropriate risk prevention strategies can be implemented for the future. Such logic seems unassailable: the idea of getting to the bottom of something is the essence of the can-do American entrepreneurial spirit and a little-questioned aspect of our postindustrial culture. Find the problem and fix it. Studies in risk analysis, however, suggest that this is not always possible. Yale sociologist Charles Perrow believes that accidents in high-risk technologies—chemical plants, nuclear power plants, ships, traffic control systems, dams, nuclear weapons, space missions, and aircraft — may simply have no clear cause. He believes that a new type of catastrophe has emerged, a new breed of mega-accident that has occurred only during the last fifty years and mostly during the past quarter century. His thesis is both frightening and compelling.

While accidents are traditionally the product of a human error, design flaw, or procedural problem, this new type of accident is the result of the very complexity of the technological systems themselves. Complex technological systems such as those in a chemical plant require the successful interaction of many parts. Small failures occur routinely. Perrow argues that these small failures will periodically combine in unforeseen ways to create much larger failures, which can shut down the entire operation and lead to catastrophe. Even with the most effective safety devices, the occurrence of accidents is the inevitable price of highly complex technology. He dubs this new breed of problem "normal accidents." They're normal not because they are frequent, but because they are inevitable given the normal workings of a complex system. Every highly complex system, such as a nuclear power plant, has thousands or more "components," each capable of sustaining numerous small failures, whether it's a broken gauge, a stuck valve, a tired technician, or a faulty procedure. When two or more of these failures interact in an unexpected way, the scale of the accidents can escalate.

"Many a mickle make a muckle,' goes an old Scottish saying, meaning that many little things can add up to something big. Many accidents occur not because of massive pipes suddenly splitting, wings falling off, or thunderous explosions. Instead, they're caused by a string of seemingly insignificant failures that cumulatively cause something much larger than the sum of their parts.

One day this past winter, technological failure in a tightly linked system prevented me from getting to my daughter's gymnastic debut. This event had been anticipated in our household for months. My wife had planned to come home early from a conference in Chicago in time to make our daughter's show. I had cleared my desk of all writing assignments so that I could definitely attend. Early that morning, I made sure that she had packed her leotard and kissed her goodbye. I gave her the last ten dollars in my pocket for flowers for her teacher. Moments later, I received a call from an editor. He was in a bind, he said. A writer assigned to write a short piece had just lost his father and was unable to finish the assignment. Of the writers he knew, I was the only one who knew the subject well enough to finish the story in time for the deadline early that afternoon. "Please," he added. I agreed reluctantly, mostly because he was a friend. Loaded with coffee, I quickly wrote the piece and faxed it off. If I hurried, I calculated, I could still just make the show.

On my way out the door, the phone rang. My wife was on the line, reporting that her plane had never left Chicago because of a snowstorm. She couldn't possibly make the show—all the more important, then, that I get there. I rushed out the door, jumped in the car, and turned the key in the ignition, only to hear a disturbing clicking noise. The ignition system had died in the cold weather. Okay, I thought, I'll take the Honda. The Honda started up, but I realized that I was running on fumes (our babysitter had forgotten to tell me the car was nearly out of gas). I drove quickly into town to the gas station, only to get caught in traffic caused by an accident. I sped off down a side street to another gas station I knew, just sputtering to a pump before the engine conked out. It turned out that this gas station wouldn't take any of my credit cards, and I had given my cash to my daughter. The attendants told me there was an ATM several blocks away. I parked and locked the car and ran to the machine, only to find that it was being serviced. With the subway only a few more blocks away, I sprinted there, figuring I could catch a train to the show. I always keep a subway card in my wallet as a backup. I raced to the station entrance, only to find it closed because someone had fallen down the escalators. Amazingly, I managed to borrow some money from an acquaintance passing by but couldn't find a cab because other thwarted subway riders had already occupied them all. By that time, the show had started and I'd missed my daughter's performance.

When I saw my daughter that evening and confronted the look of betrayal in her eyes, what should I have told her? What was the reason for my failure to get to her show? That I ran late (a procedural problem)? That the car didn't start (mechanical failure)? That the babysitter had forgotten to fill up the car (human error)? That the ATM was being serviced and the subway station was closed (circumstance)? None of these reasons alone explains why I didn't make it. It was the cumulative interaction of these small failures that conspired to frustrate my intentions.

While this example is a little far-fetched, it does illustrate how happenstance that morning linked obstacles that I couldn't possibly have anticipated or imagined: the death of a writer's father and a car's failure to start. A snowstorm in Chicago and a closed subway station. Even the safety backups — my wife, a second car, a subway ticket in my wallet — all failed, and I couldn't achieve my objective. Any of these mishaps alone might have constituted an annoyance, but none of them individually would have prevented me from getting to the show on time. I still don't have an adequate explanation for my daughter.

Such an unlikely combination of small events contributed to the near disaster that occurred in March 1979 at the Three Mile Island nuclear power plant in Harrisburg, Pennsylvania. Except this time, the result was not just a father missing his daughter's show but the near meltdown of a nuclear reactor core. The momentousness of this potential catastrophe cries out for an equally dramatic cause — a heinous terrorist plot or the rampant greed of a corporation. The train of events, however, appears disappointingly pedestrian. It began with the water polisher, a device that filters and purifies water for the plant. The polisher was prone to breakdowns, which normally didn't prove to be a problem. Except this time. A cupful of water crept through a faulty seal into the plant's air system. Two water pumps automatically shut down as a result, because the moisture appeared to interrupt air pressure to two valves. The absence of these pumps meant that cold water no longer flowed into the steam generator.

Engineers, in designing the system, had foreseen the possibility of the failure of the water pumps, so emergency backup pumps started to operate. However, the valves to these pumps happened to be closed, so the water still wasn't being pumped. As fate would have it, a repair tag obscured the dial in the control room (one of hundreds) indicating that these valves were closed, not open. To add insult to injury, a final backup safety system known as an "electromatic relief valve" also malfunctioned, sticking open when it should have been closed. Technicians in the control room didn't learn that the relief valve was stuck open, because a faulty position indicator indicated that it had shut. They knew something was amiss overall, but it took precious time to learn the cause. In the meantime, the reactor came close to a meltdown.

In hindsight, the chain of events seems quite evident, as it does when Sherlock Holmes's deductions and reasoning are laid out at the end of one of his cases. No one could have foreseen the series of events that occurred at that in place at Three Mile Island. Imagining the possible interactions and failures in a complex system such as that in place at a nuclear power plant involves an infinite number of possibilities. This calls into question whether the most advanced safety engineers could ever "manage" the risks in such a highly complex technological system. Fortunately, most of what occurs in our lives, in nature, and in human-made systems occurs in a straightforward, linear fashion. Most of the time, when I want to go somewhere, I get into my car and drive there, plain and simple. The events in a nuclear power plant most often move along a planned path. However, complexity always lies in wait around the corner, and sometimes events change radically enough to cause a missed performance or a technological catastrophe. These "complex interactions," says Perrow, "are full of branching paths, feedback loops, jumps from one linear sequence to another." Uncomfortable as it may seem, the technological complexity in modern society forces us away from the assumption that science can work toward zero risk.

A particularly nasty and paradoxical case of relative risk and technological tragedy seems to have occurred with the crash of ValuJet Flight 592 in May 1996. Investigators finally pieced together the information that the jet's front hold contained a box of discarded but undischarged oxygen canisters. Sparks caused these canisters to explode, sending the twin-engine DC-9 into the Everglades with the loss of all 110 passengers and crew members. The irony is that the oxygen generators are steel canisters that fuel the oxygen masks that drop from the ceiling in cases in which the cabin depressurizes; essentially, they are risk-management devices. Of course, the canisters should have been discharged and decommissioned properly. Pilot and journalist William Langewiesche writes that "ValuJet Flight 592 burned and crashed not because the airplane failed but, in large part, because the airline did." A series of minor mishaps and procedural confusion caused the oxygen generators to be loaded onto Flight 592 while still containing oxygen and heat-producing chemicals.

Boston College sociologist Diane Vaughn describes how in another incident, NASA and a manufacturing contractor involved in the space shuttle program gradually and incrementally stretched the range of acceptable risk, until the entire operation was jeopardized. She calls this the "normalization of deviance." The end result was the failure of the O-rings and the *Challenger* tragedy. "No fundamental decision was made at NASA to do evil," she writes, "Rather, a series of seemingly harmless decisions were made that incrementally moved the space agency toward a catastrophic outcome." According to these students of "system accidents," nuclear power plants, major dams, chemical plants, and aircraft traffic control systems will inevitably face accidents at unpredictable times.

Some experts believe that learning what went wrong and building new risk-management devices to prevent recurrences in the future will in fact not achieve that goal. It's not clear that lessons learned from Three Mile Island could stop similar disasters from happening again. Paradoxically, additional redundant safety features — an extra set of switches or safety procedures, for instance — or even extra risk-management devices — those oxygen canisters — could add to the probability that a "normal accident" could occur because there would be even more parts that could malfunction. What, then, are the costs of air travel, space shots, nuclear power, and chemical plants? Could they be the inevitability of catastrophe every now and then? If the public insists on more frequent and cheaper air travel, space exploration, inexpensive power, and a wider range of chemical products, the answer appears to be yes. In most cases, except nuclear power plants, the American public has tacitly agreed that the benefits are worth the risks.

Over the past two decades, scientists have puzzled over the nature of complexity more generally. Looking at regularly complex systems — such as the weather and the stock market — they've searched for commonalities in systems that seem to share only one thing, namely, their complexity. Chaos theory is a part of the science of complexity. A famous apocryphal example is that of a butterfly flapping its wings in China that displaces a small pocket of air, which eventually causes a hurricane to hit North America. It suggests that in nonlinear, dynamic, complex systems, seemingly insignificant events can have huge impacts down the line — or, more often, none at all.

The ideas surfacing within complexity theory throw a wrench into the intellectual machinery that has been in operation ever since Sir Isaac Newton asserted that "what is describable is predictable."

The Newtonian world assumes that there is an inherent order in the world that scientists can figure out if they dig deeply enough. Complexity theory suggests that even if science can describe a process or system, though, it can't necessarily predict the course of events in that process. That's why weather reports and stock analysts' predictions are notoriously untrustworthy. But what does this have to do with risk? If our world contains systems that are not predictable, then we cannot determine the probabilities of risk. Risk analysis has very real limitations.

Even if the outcome of a complex system can't be predicted, computers can model the consequences of various scenarios. Computer modeling approaches the question of risk obliquely, much as Blaise Pascal did when he pondered the consequences of believing in God more than three hundred years ago. He avoided the trap of trying to answer the age-old question, instead examining the consequences of believing or not believing in God. In the same way, computer modeling may not be able to predict the likelihood of a particular event occurring in a complex system, but it can examine the consequences of an accident or failed system. By juggling thousands of variables, computer models can play out a range of consequences. This can prove useful in risk management because it can reveal which safety systems or risk-management strategies are better than others.

One day I witnessed the power of computer modeling in San Francisco. On a screen, I watched as flames engulfed the Bay Area after a large earthquake. Fanned by strong winds from the west, fire raced across the city. The computer counted the minutes as fire engines moved along a gridwork of streets to battle the inferno. . . . Thus has Hemant Shah of Risk Management Solutions re-created the conditions of the 1906 San Francisco earthquake and overlaid it onto today's city. The company performs this simulation to anticipate what might happen in the event of a quake. Factored into this model are countless specifics that only a computer could juggle: the location of gas mains, the kinds of firebreaks, the speed of fire engines battling rubble in the roadways, the type of construction of various buildings, as well as soil types and ground conditions.

Seven hours later, when the fires have burned out or been controlled, a series of blimp- and cigar-shaped splotches mar the city grid, indicating fire damage. With this information, Shah can calculate the claim costs for an insurance company should a similar earthquake occur again. Much useful information could be generated by running this model. Local fire officials could look at the response times of their teams. If certain bridges were likely to go down in a fire, alternative routes could be assigned ahead of time so that traffic would not grind to a halt. Zones of housing susceptible to earthquake damage due to their location could be reinforced and residents trained in rapid evacuation.

Of course, differences in wind conditions, season, and strength of earthquake obviously exert a major influence over the shape of the disaster. And the computer programmer must make myriad assumptions about how fast fire spreads. However, modeling an earthquake in a highly populated area or another catastrophic event does enable technicians to begin visualizing the consequences of disasters. Perhaps in these admittedly theoretical exercises some risk-management strategies may be gleaned. But then, society and individuals will face more choices and decisions. Knowing that they live in a house likely to be destroyed in a moderate earthquake, would a family pay higher premiums, or would they move away? How much money is a community, the state, or nation willing to devote to emergency management to reduce the harmful aftereffects of an earthquake?

All of us, individually as well as collectively, are forced more and more to manage risk and our fate. Doing so involves shifting our perspective away from the notion of eradicating risks from our lives one by one, abandoning the idea that zero risk is somehow reachable. Instead, the pillar of the modern language of risk is that risk itself is a fluid, not static, concept. Our successes in fighting risk have themselves created new orders of risk for us to deal with. In light of this, we must learn to manage our risks, picking and choosing, and negotiating, not just responding to them. What, we must ask ourselves, are the risk-related ramifications of killing that snake in the woods outside our house?

CHAPTER TEN

OUR NEW ROLE AS RISK MANAGERS

For centuries, we've placed our trust in experts. We still do, of course. However, of late the changing risk landscape has altered the nature of that trust. As notions of risk have changed, and we've pulled our fate away from the gods and into our own hands, the relationship between expert and layperson has crossed into uncertain terrain.

When I questioned our obstetrician that winter morning years ago about her automatic decision that my pregnant wife needn't undergo an invasive genetic test, her response was to draw back and become defensive. After a cursory discussion of the risks and benefits of the test in our situation, she wouldn't say more. I remember her terse and final response, "Well then, I can't advise you." The palpable fear on her face betrayed concerns that our emotions might suddenly erupt to the surface like a volcano, or that something she might say would make her vulnerable to a lawsuit or other problems in the future.

Clearly, this doctor, although a highly trained and competent caregiver, had neither the interest nor the tools to help my wife and me work through this risk decision. We were left either to accept her decision or ponder the ramifications of one choice or another by ourselves.

This decision was as real a risk as an expedition faces when it confronts a yawning abyss. The pivotal issue involved a tricky evaluation of the risks of finding out information versus not. Amniocentesis, a procedure in which a physician samples a woman's amniotic fluid by inserting a large-gauge needle into her abdomen, can determine whether the fetus has certain genetic disorders, such as Down syndrome. Bringing up a child with the special physical, intellectual, and emotional needs resulting from Down syndrome involves many sacrifices and joys not associated with raising a healthy child.

The information comes at the cost of some increased risk, however: about 1 out of every 200 procedures causes the woman to lose the fetus. The situation becomes tricky as a woman grows older, because the risk of having a Down syndrome child goes up with the age of the mother, as do the difficulties of getting pregnant and carrying to term. Get the test, and you risk losing the fetus; don't, and you can't know if the fetus has a genetic abnormality, so you forfeit the chance to act on that information, either by having an abortion or preparing emotionally and practically for the arrival of a child with special needs.

To help doctors and parents decide when the risk of having a Down syndrome child is high enough to warrant the risk of the test, the medical community has established the recommendation that any healthy, pregnant woman without complicating factors in family history should undergo an amniocentesis if she's thirty-five years or older. Our doctor held this up as fact, not a recommendation.

In making the decision for us that no test was needed — my wife wasn't yet thirty-five — she misled us. No one would argue that a 1 in 365 chance of bearing a child with Down syndrome at age thirty-five is insignificant. Yet the number is essentially arbitrary, because there's no one universal number that establishes whether the probability of an event occurring is significant enough to act upon. All potential parents bring to this risk decision a complex stew of personal beliefs, values, medical issues, economic resources, lifestyle expectations, and emotions. But these factors differ so much from one set of potential parents to another that they could alter that benchmark substantially, shifting it years higher or lower. What one couple considers significant may be a trifle to another. Far from being set in stone, the benchmark is only a guidepost, a piece of information with which to start the decision-making process.

One set of parents we know placed a high premium on not having a child with Down syndrome. Their decision was for the wife to have amniocentesis at age thirty. Another couple we know in their late thirties would never consider having an abortion even if they learned that their child had a genetic abnormality. For them, taking the test made little sense. Other couples, like my wife and me, fall into a gray area. Our values led us to believe that getting pregnant entailed certain responsibilities and incumbent risks. On the other hand, we wanted to be pragmatic about what we could offer a child with severe disabilities. During our brief discussion with our physician, I'm certain that she had no idea about our beliefs or values regarding all these issues. To her, the benchmark of age thirty-five held all the information necessary for this decision.

I felt the doctor's representation of the benchmark as all we needed to know was misleading and dangerous because it ignored our individual circumstances by not providing us with sufficient information to judge the risk ourselves. As a society, of course, we do need guidelines to avoid endless indecision. Without such benchmarks as the FDA's dietary guidelines (found on mandatory food labels), drug and vaccine doses, and pollution standards, society would be in a perpetual state of confusion. In our reliance on these guidelines and benchmarks, however, we often forget that they represent only best guesses, based on averages and many assumptions. Setting guidelines we can live with involves carefully blending the best available science, agreeing on our tolerance for risk, and clearly delineating what we hold most important. While experts such as health care providers can give us solid recommendations upon which we must rely, the decision ultimately lies in our lap, not theirs. These are personal questions that only we ourselves can answer.

Critical to understanding benchmarks is the idea that dropping or lowering them has its own implications. Mandating that superfund sites must be cleaned up entirely, to the point where they present zero risk, makes emotional sense, but it may not make good risk sense. One EPA report defined a clean site as one where a child could eat the dirt at the site more than 200 days a year without harm. Going to such extremes to banish risk by pursuing the zero-risk holy grail in such a manner will exert tremendous impact on other risk-management situations; for instance, fewer resources will then be available to fight other health threats to children. Perhaps the children would be better served by better-quality day care that would prevent them from eating that dirt in the first place.

Or consider nuclear energy for consumer power and industry in this country. This source of energy does not enjoy much support, in large part because the technology pushes many of our unconscious risk buttons. In our new role as risk managers, we must look beyond the traditional perspective of crushing risks such as nuclear power one by one and focus on a larger picture instead. Not using nuclear power presents trade-offs. We often forget that burning fossil fuels creates substantial hazards of its own: possible global warming, poisonous emissions into the atmosphere, strip mining and disruption of pristine natural areas, accidents and disease among coal miners and during the transportation of materials, and imbalances among world powers. We may still decide that nuclear fission power as an energy source is not worthwhile, but it must come in the declared context of other risk trade-offs.

Fortunately, risk analysis, rooted in the emergence of probability and statistics over the past several centuries, gives us the tools to evaluate these risk decisions effectively. Fluency in the language of risk doesn't require an individual to crunch algorithms, work out complex problems in probability theory, or memorize vast quantities of risk ratios. Instead, the language of risk gives us a larger context in which to think about, frame, and evaluate risk. The elements of the language include the concept of relative risk and the idea of risk trade-offs; the knowledge that risk lurks wherever there is uncertainty; a working familiarity with risk ratios, time scales, and averages; the ability to control basic risk misconceptions, emotions, and unconscious biases; and the knowledge that individuals carry different biological predispositions toward risk. Using these tools enables us to avoid the confusion and traps of old risk responses, in which we responded to risks today with potentially dangerous simple avoidance or search-and-destroy strategies. We have the opportunity now to become managers who can select among various consequences of our actions, work with risks, and allocate scant resources judiciously for maximum effect.

We will need these tools to cope with the complex problems of the new century. What do we do when toxicologists, using increasingly sophisticated quantitative measuring devices, discover traces of toxins in our homes, food, and workplace? How do we act on new information about how artificial hormones affect the environment? Risk is at the heart of deciding what is natural and what is not. Should people involuntarily exposed to a carcinogen that might cause cancer in a decade or two receive compensation for sickness and disease not yet manifested in any symptoms? As certain risk issues become more clearly delineated, should we make risk adjustments and allowances for special groups, such as children and the economically disadvantaged?

The new century will require us to navigate through complex risk decisions involving ethics and personal freedoms: questions about fertility drugs and women conceiving well past the normal age of menopause; cloning and biomedical engineering; the ethical and practical issues that will continue to emerge as molecular biologists unearth more secrets about our DNA; and how to balance trade-offs between material benefits and the degradation of the environment. Where do we set state and federal speed limits that balance convenience with safety? How much more are we willing to pay in taxes on gasoline to carry out an environmental cleanup that will benefit future generations?

Increasingly, difficult and frightening personal decisions will fall into our laps. Should a young woman undergo a radical mastectomy before any signs of cancer surface if she learns that her genes and family history dispose her to an 85 percent chance of contracting breast cancer at some point in her life? Do we opt for an expensive drug regimen that extends life but results in persistent pain, nausea, and distress? Is estrogen treatment worth the protection against hip fractures when it slightly raises a woman's chances of contracting cancer? We will make some of these risk decisions in the doctor's office, at the supermarket, in the boardroom, in heated family discussions at the kitchen table, and in the courtroom. We will make others with our checkbooks, protesting certain actions with letters or walkouts, and voting for particular candidates, agendas, and referendums.

The choice remains ours. We can decide not to take part in risk decisions in our lives. We can act as my expedition did when confronted with the possibilities of encountering a polar bear in the Arctic. We let disagreement and misleading notions about risk maneuver us into not adopting any kind of strategy. By not working out a rational plan, we left ourselves vulnerable, placing our safety in the hands of fate. We happened to be lucky that time. We can choose to remain in a state of ignorance and blame the media for presenting what seems to be conflicting information. We can resort to old-fashioned notions of risk, rationalizing that "what was good for my grandfather is good enough for me." Or, instead, we can take positive steps to educate ourselves in the language of risk and take ongoing responsibility for analyzing hard questions, making difficult choices, and reevaluating those decisions as necessary in the face of new information.

Researching this book has taught me that the process of working through risk information is as important as coming up with the final answer. In the case of the amniocentesis test, my wife and I could have blindly accepted the doctor's decision. The decision we finally made — not to undergo the test — did agree with the recommendation that our doctor had made. However, it became our decision, not hers, because we went through the process of defining and elucidating what was important to us. It's much easier to accept the doctor's decision, or that of any expert for that matter, as gospel rather than go through the laborious and often painful process of evaluating the risk information ourselves. Yet it is the latter that we must take if we want to become good risk managers. Good risk management results in better choices and outcomes for us to live and work with. Going through the evaluation of a risk decision will also prepare us for those cases, increasingly common in modern times, where there are no guidelines and the information available is conflicting.

In the face of an intimidating health professional or an authoritative statement from an expert, it's all too easy to forget that decisions ultimately remain ours to make. Most risk decisions we make as individuals and collectively as a society have no right or wrong answers. The decisions we face — to smoke or not, to save one species or another — emerge not as matters of right or wrong, but rather as judgment calls involving serious trade-offs between subjective understandings of what's good and bad. That's why we can't rely on a bunch of experts to decide for us. Now we are responsible not only for our own bodies and communities but also for the Earth itself.

If we Americans don't learn the language of risk, we face a bleak future indeed. We will surrender, become apathetic, and look to quacks and "new age" cures for guidance. The tort system, already overwhelmed with too many frivolous lawsuits, will become even more crowded as Americans fail to take responsibility for life around them and cast about for someone to blame. Operating under the fallacy that the world has grown too complex for ordinary people to understand, experts will make more and more of the decisions that people should make for themselves. For every decision ceded without explanation to an expert, Americans lose a small piece of freedom and give up a little more control over their lives. As Thomas Jefferson wrote: "If we think they [the people] are not enlightened enough to exercise their control with a wholesome discretion, the remedy is not to take it from them, but to inform their discretion." When experts lament, "How can the public understand complex issues?" our response must be, "How can we afford not to?"

The public must remain active and vigilant. We must demand reliable information about risks and a context for understanding their significance and relevance to our lives. It's incumbent upon us to ask legislators and other individuals in positions of trust and authority to spell out the specific trade-offs when regulations and risk-management strategies are being considered.

The public must get better at recognizing when special interest groups, advertisers, or tabloid journalists inflate risks by appealing to our fears and emotions. Americans must insist that journalists routinely use numeric risk ratios and always present other examples of equally probable risks in everyday life so that we don't have to guess at the real severity of the risks being discussed.

We must require health care workers to present risk options in clear and understandable terms. Perhaps most important, we must realize that we can't expect scientists to deliver definitive answers to many of the difficult questions we face. In researching this book, I learned that although I'm an environmentalist, I still want to drive a car and enjoy a high standard of living. Yet I realized that I am willing to reject some of what that standard requires for the sake of the health of my children, their children, and the protection of the environment.

Speaking the language of risk has heightened my disdain for companies that insist on a product's safety even in the face of conclusive evidence to the contrary or for an environmental group that states apocalyptically on the basis of one small study that one substance will kill us all. I've grown weary of riding the roller coaster of industrialists, environmentalists, bureaucrats, and a range of experts who try to scare or intimidate me into buying into their agendas. Learning the language of risk has helped me chart a new path, from which I look at all the information with a grain of salt and ask tough questions. Now I want to know the context of the risk information — what's the source, how is it expressed? Just as I consult a consumer magazine when I'm in the market for a new car, I try now to think of myself as one of those editors, evaluating not a product but the merits of the risk information itself. I'm less bowled over by numbers and less disposed to believe that they impart certainty. I look at them as tools and realize that some tools do certain jobs well, others not at all.

Speaking the language of risk has kept me skeptical of early reports of health warnings: I now wait to see if further research confirms them. On the flip side, I've become suspicious of claims that a megadose of one vitamin or another or a new diet regimen will prove a risk panacea. I understand that they could well put me at greater risk. I'm less swept away by the risk information flying through the media. In fact, I've gone back to basics, relying on clear, proven strategies for maintaining health and reducing risk that appear simple but are definitely effective. I spend less time worrying about exotic new risks and more on making sure that my family eats at least five servings of vegetables and fruits every day. I've worked to lower my high blood pressure. I let other drivers cut in front of me, now conscious of all the risk involved in road rage and heightened stress. I put a higher emphasis now on getting my body in shape.

Knowing more about how to evaluate and think about risk hasn't made me more timid or risk averse. I still take risky assignments and do things that many people consider high-risk activities. However, learning the language of risk has helped me get a better grip on the true nature of the risks in my life. I have a clearer sense of my personal tolerance for risk. I'm less willing now to blame another person or institution when something goes wrong in the wake of a risk decision that I've consciously made.

At the dawn of the twentieth century, the famous American historian Henry Adams asked what would make a ideal citizen in the new century. When we ask ourselves this same question a century later, an intimate knowledge and conversance with the language of risk must rank high on our list of qualities. Ultimately, how we understand and interact with risk as individuals and as citizens is a reflection of what we care about, what we take responsibility for, and who we are. Speaking the language of risk means we will enjoy the power and the ability to shape our destinies as no people before us have ever done.

SELECTED BIBLIOGRAPHY

Ames, Bruce N., Margie Profet, and Lois Swirsky Gold. "Dietary Pesticides (99.99% all natural)." *Proceedings of the National Academy of Sciences*: vol. 87 (October 1990): 7777–7781.

Ames, Bruce N., Renae Magaw, and Lois Swirsky Gold. "Ranking Possible Carcinogenic Hazards." *Science*: vol. 236 (April 17, 1987): 271–280.

Ames, Bruce N., Mark K. Shigenaga, and Tory M. Hagen. "Oxidants, Antioxidants, and the Degenerative Diseases of Aging." *Proceedings of the National Academy of Sciences*: vol. 90 (September 1993): 7915–7922.

Anderson, Oscar E., Jr. *The Health of a Nation: Harvey W. Wiley and the Fight for Pure Food*. Chicago: University of Chicago Press, 1958.

Apter, Michael J. *The Dangerous Edge: The Psychology of Excitement*. New York: Free Press, 1992.

Bailar, John C., III, Jack Needleman, Barbara L. Berney, and J. Michael McGinnis, eds. *Assessing Risks to Health: Methodological Approaches*. Westport, Conn.: Auburn House, 1993.

Barrett, Anthony A. *Agrippina: Sex, Power, and Politics in the Early Empire*. New Haven: Yale University Press, 1996.

Benarde, Melvin A. *Our Precarious Habitat*. New York: John Wiley & Sons, 1989.

Bernstein, Peter L. *Against the Gods: The Remarkable Story of Risk*. New York: John Wiley & Sons, 1996.

Boyle, Charles, Peter Wheale, and Brian Surgess. *People, Science and Technology: A Guide to Advanced Industrial Society*. Totowa, N.J.: Barnes and Noble Books, 1984.

Braus, Patricia. "Everyday Fears." *American Demographics* (December 1994): 32–58.

Brehmer, Berndt, and Nils-Eric Sahlin, eds. *Future Risks and Risk Management*. Dordrecht, The Netherlands: Kluwer Academic, 1994.

Breyer, Stephen G. *Breaking the Vicious Circle: Toward Effective Risk Regulation*. Cambridge, Mass.: Harvard University Press, 1993.

Carson, Dennis A., and Lois Augusto. "Cancer Progression and P53." *The Lancet*: vol. 346 (October 14, 1995): 1009–1011.

Casarett, Louis J. *Casarett and Doull's Toxicology: The Basic Science of Poisons*. Edited by Mary O. Amdur, John Doull, and Curtis D. Klaassen. 4th Edition. New York: Pergamon Press, 1991.

Caufield, Catherine. *Multiple Exposures: Chronicles of the Radiation Age*. New York: Harper & Row, 1989.

Center for Risk Analysis, Harvard School of Public Health. *A Historical Perspective on Risk Assessment in the Federal Government*. March 1994.

Cipolla, Carlo M. *The Economic History of World Population*. New York: Barnes and Noble Books, 1978.

Cohen, Bernard L. "Catalog of Risks Extended and Updated." *Health Physics*: vol. 61, no. 3 (September 1991): 317–335.

———. "How to Assess the Risks You Face." *Consumers' Research* (June 1992): 11–16.

Colborn, Theo, Dianne Dumanoski, and John Peterson Myers. *Our Stolen Future: Are We Threatening Our Fertility, Intelligence, and Survival? A Scientific Detective Story*. New York: Dutton, 1996.

Conrad, J., ed. *Society, Technology, and Risk Assessment*. London: Academic Press, 1980.

Crouch, Edmond A. C., and Richard Wilson. *Risk/Benefit Analysis*. Cambridge, Mass.: Ballinger, 1982.

Cutter, Susan L. *Living with Risk: The Geography of Technological Hazards*. London: Edward Arnold, 1993.

David, F. N. *Games, Gods, and Gambling*. London: Charles Griffin & Co., 1962.

Douglas, Mary, and Aaron Wildavsky. *Risk and Culture: An Essay on the Selection of Technical and Environmental Dangers*. Berkeley: University of California Press, 1982.

Duke, Richard C., David M. Ojcius, and John Ding-E Young. "Cell Suicide in Health and Disease." *Scientific American* (December 1996): 80–87.

Dunn, Judy, and Robert Plomin. *Separate Lives: Why Siblings Are So Different*. New York: Basic Books, 1990.

Evans, Leonard. *Traffic Safety and the Driver*. New York: Van Nostrand Reinhold, 1991.

Eysenck, H. J. *The Biological Basis of Personality*. Springfield, Ill.: Charles C. Thomas, 1967.

Farley, Frank. "The Type T Personality." In *Self-Regulatory Behavior and Risk Taking: Causes and Consequences*, edited by L. L. Lipsitt and L. L. Mitnick. Norwood, N.J.: Ablex, 1991.

Fischhoff, Baruch. "Ranking Risks." *Risk: Health, Safety & Environment* (Summer 1995): 191–202.

Fischhoff, Baruch, Ann Bostrom, and Marilyn Jacobs Quadrel.

"Risk Perception and Communication." *Annual Review of Public Health*: vol. 14 (1993): 183–203.

Flower, Raymond, and Michael Wynn Jones. *Lloyd's of London: An Illustrated History*. New York: Hastings House, 1974.

Forrest, D. W. *Francis Galton: The Life and Work of a Victorian Genius*. London: Paul Elek, 1974.

Galton, Francis. *Memories of My Life*. New York: E. P. Dutton & Company, 1909.

Glickman, Theodore S., and Michael Gough, eds. *Readings in Risk*. Washington, D.C.: Resources for the Future, 1990.

Gold, Lois Swirsky, Thomas H. Slone, Bonnie R. Stern, Neela B. Manley, and Bruce N. Ames. "Rodent Carcinogens: Setting Priorities." *Science*: vol. 258 (October 9, 1992): 261–265.

Gordon, D. W., G. Rosenthal, J. Hart, et al. "Herbal Hepatotoxicity." *Journal of the American Medical Association*: vol. 273 (1995): 489.

Gould, Leroy C., Gerald T. Gardner, Donald R. DeLuca, Adrian R. Tiemann, Leonard W. Doob, and Jan A. J. Stolwijk. *Perceptions of Technological Risks and Benefits*. New York: Russell Sage Foundation, 1988.

Hacking, Ian. *The Emergence of Probability*. Cambridge: Cambridge University Press, 1975.

Harper, Andrew C., and Laurie J. Lambert. *The Health of Populations*. New York: Springer, 1994.

Harris, Curtis C. "p53: At the Crossroads of Molecular Carcinogenesis and Risk Assessment." *Science*: vol. 262 (December 24, 1993): 1980–1981.

Hayflick, Leonard. *How and Why We Age*. New York: Ballantine Books, 1994.

Herrnstein, Richard J., and Charles Murray. *The Bell Curve:*

Intelligence and Class Structure in American Life. New York: Free Press, 1994.

Hoel, David G., Richard A. Merrill, and Frederica P. Perera. "Risk Quantification and Regulatory Policy." *Cold Springs Harbor Laboratory*, Banbury Report 19 (1985).

Holland, Heinrich D., and Ulrich Petersen. *Living Dangerously: The Earth, Its Resources, and the Environment*. Princeton, N.J.: Princeton University Press, 1995.

Horvath, Paula, and Marvin Zuckerman. "Sensation Seeking, Risk Appraisal, and Risky Behaviour." *Personality and Individual Differences*: vol. 14, no. 1 (1993): 41–52.

Hulka, Barbara S., Timothy C. Wilcosky, and Jack D. Griffith. *Biological Markers in Epidemiology*. New York: Oxford University Press, 1990.

Karlen, Arno. *Man and Microbes: Disease and Plagues in History and Modern Times*. New York: G. P. Putnam's Sons, 1995.

Kates, Robert W., Christopher Hohenemser, and Jeanne X. Kasperson, eds. *Perilous Progress: Managing of the Hazards of Technology*. Boulder, Colo.: Westview Press, 1985.

Keeney, Ralph L. "Sounding Board Decision About Life-Threatening Risks." *The New England Journal of Medicine*: vol. 331, no. 1 (July 21, 1994): 193–196.

Kevles, Daniel. *In the Name of Eugenics: Genetics and the Uses of Human Heredity*. New York: Alfred A. Knopf, 1985.

Koren, Herman. *Handbook of Environmental Health and Safety: Principles and Practices*. Volume I, 2nd ed. Chelsea, Mich.: Lewis Publishers, 1991.

Kunreuther, Howard, and Paul Slovic, special editors. "Challenges in Risk Assessment and Risk Management," in *The Annals of the American Academy of Political and Social Science*, May 1996.

Lagadec, Patrick. *Major Technological Risk: An Assessment of Industrial Disasters*. Oxford: Pergamon Press, 1982.

Lappé, Marc. *Chemical Deception: The Toxic Threat to Health and the Environment*. San Francisco: Sierra Club Books, 1991.

Laudan, Larry. *The Book of Risks: Fascinating Facts About the Chances We Take Every Day*. New York: John Wiley & Sons, 1994.

Leffell, David J., and Douglas E. Brash. "Sunlight and Skin Cancer." *Scientific American* 275 (July 1996): 52–59.

Lewis, H. L. *Technological Risk*. New York: W. W. Norton & Company, 1990.

Lillie, David W. *Our Radiant World*. Ames: Iowa State University Press, 1986.

Lindley, D. V. *Making Decisions*. London: John Wiley & Sons, 1985.

Lipsitt, L. P. and L. L. Mitnick, eds. *Self-Regulatory Behavior and Risk Taking: Causes and Consequences*. Norwood, N.J.: Ablex, 1991.

Loehlin, John C. *Genes and Environment in Personality Development*. Newbury Park: Sage, 1992.

Medina, John J. *The Clock of Ages: Why We Age — How We Age — Winding Back the Clock*. Cambridge: Cambridge University Press, 1996.

Motz, Lloyd, and Jefferson Hane Weaver. *The Story of Mathematics*. New York: Avon Books, 1993.

Muller, James E., MD, et al. "Triggering Myocardial Infarction by Sexual Activity." *Journal of the American Medical Association*: vol. 275 (May 8, 1996): 1405–1409.

National Research Council. Committee on Risk Assessment Methodology, Board on Environmental Studies and Toxicology, Commission on Life Sciences. *Issues in Risk Assessment*. Washington, D.C.: National Academy Press, 1993.

———. Committee on Comparative Toxicity of Naturally Occurring Carcinogens, Board on Environmental Studies and Toxicology, Commission on Life Sciences. *Carcinogens and Anti-Carcinogens in the Human Diet: A Comparison of Naturally Occurring and Synthetic Substances*. Washington, D.C.: National Academy Press, 1996.

National Safety Council. *Accident Facts*. 1997 Edition. Itasca, Ill.

Nelkin, Dorothy, ed. *Controversy: Politics of Technical Decisions*. Beverly Hills: Sage, 1979.

Oldstone, Michael B. A. *Viruses, Plagues, and History*. New York: Oxford University Press, 1998.

Ott, Wayne R., and John W. Roberts. "Everyday Exposure to Toxic Pollutants." *Scientific American* (February 1998): 86–91.

Ottoboni, M. Alice. *The Dose Makes the Poison*. Berkeley: Vincente Books, 1984.

Paulos, John Allen. *Innumeracy: Mathematical Illiteracy and Its Consequences*. New York: Hill and Wang, 1988.

Pearson, Karl. "The History of Statistics in the 17th and 18th Centuries Against the Changing Background of Intellectual, Scientific and Religious Thought." In *Lectures by Karl Pearson Given at University College London During the Academic Sessions 1921–1933*. Edited by E. S. Pearson. New York: Macmillan, 1978.

Perrow, Charles. *Normal Accidents: Living with High-Risk Technologies*. New York: Basic Books, 1984.

Pochin, Edward. *Nuclear Radiation: Risks and Benefits*. Oxford: Clarendon Press, 1983.

Presidential/Congressional Commission on Risk Assessment and Risk Management. *Risk Assessment and Risk Management in Regulatory Decision-Making*. Final report, vol. 2, 1997. Washington, D.C.: Government Printing Office.

Richardson, Mervyn, ed. *Toxic Hazard Assessment of Chemicals*. Burlington House, London: The Royal Society of Chemistry, 1986.

Rodricks, Joseph V. *Calculated Risks: Understanding the Toxicity and Human Health Risks of Chemicals in Our Environment*. Cambridge: Cambridge University Press, 1992.

Rothman, Alexander J., William M. Klein, and Neil D. Weinstein. "Absolute and Relative Biases in Estimations of Personal Risk." *Journal of Applied Social Psychology*: vol. 26, no. 14 (1996): 1213–1236.

Rowe, William D. *An Anatomy of Risk*. New York: John Wiley & Sons, 1977.

Schwing, Richard C., and Dana B. Kamerud. "The Distribution of Risks: Vehicle Occupant Fatalities and Time of the Week." *Risk Analysis*: vol. 8, no. 1 (1988): 127–133.

Schulte, Paul A., and Frederica P. Perera, eds. *Molecular Epidemiology: Principles and Practices*. San Diego: Academic Press, 1993.

Shubik, Martin, ed. *Risk, Organization, and Society*. Boston: Kluwer Academic, 1991.

Slovic, Paul, Baruch Fischhoff, and Sarah Lichtenstein. "Rating the Risks." *Environment*: vol. 21, no. 3 (April 1979): 14–20, 36–39.

Sprent, Peter. *Taking Risks: The Science of Uncertainty*. London: Penguin Books, 1988.

Stigler, Stephen M. "Francis Galton's Account of the Invention of Correlation." *Statistical Science*: vol. 4, no. 2 (May 1989): 73–86.

———. *The History of Statistics: The Measurement of Uncertainty Before 1900*. Cambridge, Mass.: The Belknap Press of Harvard University Press, 1986.

Sulloway, Frank J. *Born to Rebel: Birth Order, Family Dynamics, and Creative Lives*. New York: Pantheon Books, 1996.

Taubes, Gary. "Epidemiology Faces Its Limits: A Special News Report." *Science*: vol. 269 (July 14, 1995): 164–169.

Tengs, Tammy O., et al. "Five-Hundred Life-Saving Interventions and Their Cost-Effectiveness." *Risk Analysis*: vol. 15, no. 3 (1995): 369–384.

Tversky, Amos, and Daniel Kahneman. "Rational Choice and the Framing of Decisions." *Journal of Business*: vol. 59, no. 4 (1986): 251–278.

Taylor, Michael Ray. "Deep, Dark, and Deadly." *Sports Illustrated*: vol. 81 (October 3, 1994): 5–12.

Urquhart, John. *Risk Watch: The Odds of Life*. New York: Facts on File Publications, 1984.

U. S. Congress, Office of Technology Assessment. *Researching Health Risks*. Superintendant of Documents, U.S. G.P.O., distributor, 1993.

U. S. Congress, Presidential/Congressional Commission on Risk Assessment and Risk Management. Washington, D.C.: Commission on Risk Assessment and Risk Management, 1997.

Vaughan, Diane. *The Challenger Launch Decision: Risky*

Technology, Culture, and Deviancy at NASA. Chicago: University of Chicago Press, 1996.

Walter, Gerard. *Nero*. London: George Allen & Unwin, 1957.

Wehrwein, Peter. "Preventing Cancer: Halfway to Victory?" *Harvard Public Health Review*: (Fall 1996): 8–15.

Wertheim, Alfred H. *The Natural Poisons in Natural Foods*. Secaucus, N.J.: Lyle Stuart, 1974.

Whipple, Chris, ed. *De Minimus Risk*. New York: Plenum Press, 1987.

Wildavsky, Aaron. *Searching for Safety*. New Brunswick: Transaction Books, 1988.

Wilde, Gerald J. S. *Target Risk*. Toronto: PDE Publications, 1994.

Wilson, Richard. "Analyzing the Daily Risks of Life." *Technology Review*: vol. 81, no. 4 (February 1979): 41–46.

Ziegler, A., A. S. Jonason, D. J. Leffell, J. A. Simon, H. W. Sharma, J. Kimmelman, L. Remington, T. Jacks, and D. E. Brash. "Sunburn and p53 in the Onset of Skin Cancer." *Nature* 372 (December 22–29, 1994): 773–6.

Zuckerman, Marvin. *Behavioral Expression and Biosocial Bases of Sensation*. Cambridge: Cambridge University Press, 1994.

——. "Good and Bad Humors: Biochemical Bases of Personality and Its Disorders." *Psychological Science*: vol. 6, no. 6 (November 1995): 325–332.

——. "Sensation Seeking: The Balance Between Risk and Reward." In *Self-Regulatory Behaviour and Risk Taking: Causes and Consequences*, edited by L. P. Lipsitt and L. L. Mitnick. Norwood, N.J.: Ablex, 1991.

INDEX

Accident Facts, 42
Accidents
  automobile, 40 – 41, 45
  bicycle, 49
  household, 36 – 37, 45 – 46, 49
  incidence of, 50
  mega-, 158
  perceptions about, 74 – 75
Acer, David, 75
Aconite, 125
Actuarial tables, 17
  precursors of, 22
Adams, Henry, 176
Aerosols, effects of, 156
Aflatoxin B1, 117, 129
Agent Orange, 131, 139 – 140
Agriculture, development of, 56
Air bag
  benefits of, 41, 150
  costs of, 150 – 151
Alar, 77 – 79, 131
Alcohol
  neurotransmitters and, 99
  risks and benefits of, 151
  risks from, 61, 151
Allyl isothiocyanate, 126
Altai mountains, 105
Amanita phalloides mushroom, 125
Amanitin, 125
Amber, 126
Ames, Bruce, 111, 123 – 127, 145 – 146
Aminotriazole, banning of, 63
Amniocentesis, 2, 168 – 169
Analytical geometry, 14
Animal testing, of carcinogenicity, 143 – 144
Anthocyanins, 126
Antibiotics
  resistance to, 64
  uses of, 59 – 60
Anticyclones, 25
Antilock braking systems, 81 – 82
Apoptosis, 115
Apple industry, 77 – 79
Apter, Michael, 103
Arbuthnot, John, 18
Arsenic, universal occurrence of, 128
Artificial hormones, 144 – 145
Artificial products, perceptions about, 76
Artificial sweeteners, as risk factor, 39
Asbestos, cancer risk from, 129, 151
Aspartame, 155
Aspirin, toxic dose of, 132

Atomic bomb, 154
Australian rappelling, 88, 90
Automobile accidents
  incidence of, 45
  prevention of, 41
  risk factors in, 40 – 41
Automobile
  relative safety of, 44 – 45
  weight of, 40 – 41
Auxins, 126
Averages, uses of, 22 – 23
Awakenings, 99

Basic Cave Diving, 91
Batman, 65 – 66
Becquerel, Henri, 152
Bed, risks from, 35 – 36
Behavior
  birth order and, 96 – 97
  environmental role in, 96
  twin studies of, 96
Bell curve, 23 – 25
Benzene
  as carcinogen, 118
  in tobacco smoke, 128
  universal occurrence of, 128
Benzopyrene, 113, 116
  in grilled food, 124
Bergalis, Kimberly, 75
Bernstein, Peter, 32
Bertazzi, Pier Alberto, 141
Bhopal, 156
Bicycle accidents, 49
Bill of Mortality, 20
Biomarkers, 118
Birth order, importance of, 96 – 97
Black boxes, aircraft, 157
Bonnington, Chris, 93
Borax, toxicity of, 134
Boric acid, toxicity of, 134
Botulism, 73 – 74
Bowden, Jim, 92
Boylston, Zabdiel, 57
Bradley, Ed, 77
Brain, hormones and, 98
Brash, Douglas, 116, 117
BRCA-1 gene, 119
Breyer, Stephen, 48
Britannicus, 130 – 131
Business practice, origins of modern, 18
Butter Yellow, banning of, 63

Cabbage, toxins in, 124
Cache National Forest, 87
Caffeine, toxic dose of, 132
Cancer
  cell replication in, 114
  factors correlated to, 29, 31 – 32
  genetic aspect of, 114 – 119
  incidence of, 50, 61, 114
  locations of, 129
  lung, 62
  risk assessment for, 119 – 122
  risk factors for, 151 – 152
  therapies for, 115
Carcinogens, 116 – 118
  animal testing of, 143 – 144
  measures of, 136
  naturally occurring, 123 – 129
  risk assessment of, 144
  synergistic effects of, 129
  thresholds of, 137 – 138
Carcinogens and Anticarcinogens in the Human Diet, 125
Carson, Dennis, 114
Carson, Rachel, 63
Casarett, Louis, 135
Causation, distinguished from correlation, 28 – 29, 31
Cave diving, 90 – 93
Cells
  destruction of, 111 – 112
  life of, 109
  relation of DNA to, 110 – 111
  suicide of, 115 – 116
  waste products of, 112 – 113
Chain of events, magnifying effect of, 158 – 163

Challenger space shuttle, 156, 162–163
Chaos theory, 163
Childhood diseases, conquest of, 59
Children, management of risk to, 54
Chirurgical Observations, 133
Chloracne, 140
Chloroform, 128
Cholera, conquest of, 59
Cholesterol, carcinogenicity of, 151
Chrysanthemums, 126
Cirrhosis, 61
Cocoa, adulteration of, 134
Coffee
  adulteration of, 134
  toxic dose of, 132
  toxins in, 125
Cohen, Bernard, 39
Colon cancer, risk factors for, 151
Color additives, regulation of, 136
Complexity theory, 163–166
Consumer Product Safety Commission, 63
Consumer products, number of, 155
Copernicus, Nicolaus, 131
Copper, food adulteration with, 134
Copper sulfate, 133
Correlation, 28–29
Cotton sleepwear, risks from, 54, 156
Cowpox, 58
Crick, Francis, 109, 110, 118
Curie, Marie, 152, 153
Curie, Pierre, 153
Cyanide, 143
Cycloids, 11–12
Cytokinins, 126

D4 receptor gene, 99–100
Daminozide, 77
Darwin, Charles, 26
Data, proliferation of, 3, 4
Day After, The, 154
DDT
  effects of, 64
  risks and benefits of, 145
De Ratiociniis in Ludo Aleae, 18
Death angel mushroom, 125
Delaney Clause, 136, 138
  repeal of, 144
Dickens, Charles, 62
Diet, as risk factor, 61–62
Dioscorides, 131
Dioxin, 131
  accidents involving, 140–142
  regulation of, 138–139
  toxicological studies of, 139–143
  universal occurrence of, 128
Disease, perceptions of, 75
DNA
  dynamic nature of, 110–111
  effects of defects of, 114
  effects of radiation on, 113–114
  function of, 98, 110
  personality and, 97–98
  repair of, 113–114
  structure of, 63, 109–110
Domesday Book, 19
Domestication, origins of, 56
Dopamine, 99
  and extroverted behavior, 99–100
  receptor for, 99–100
Dose-response relationship, 132–133
Double-entry bookkeeping, 18
Doull, John, 135
Down syndrome, 168–169
Draft bill of exchange, 18
Dread, 74–75
Driving
  risks of, 40–41, 49
  self-assessment about, 71–72
Drowning, incidence of, 49
Dulcin, banning of, 63
Dunn, Judy, 96

Earthquake, modeling risk of, 164–165
Ebstein, Richard, 100

Endocrine disrupters, 145
Environment, role in behavior, 96
Environmental hazards, 47
  attitudes towards, 102
Environmental Protection Agency (EPA), 63
Enzymes, 98
Epidemiology, 32, 118
  molecular, 118 – 119
  origin of, 22
  risk management and, 151 – 152
Ethics, dilemmas in, 4
Ethyl carbamate, 126
Euclid, 13
Eugenics, 26
Evans, Leonard, 40, 150
Exit polls, 25
Exley, Sheck, 90 – 93
Extraversion, 94 – 95
Exxon Valdez, ship, 156
Eysenck, Hans, 94 – 95

Fallout, atomic, 154
Falls, incidence of, 49
Fat, as risk factor, 61 – 62
Fatality Analysis Reporting System, 40
Fear, 1, 90
  cultural and gender issues regarding, 102
Feed additives, regulation of, 136
Fermat, Pierre de, 14, 15
Fertilizers, and nitrogen cycle, 64
Fischhoff, Baruch, 48, 72, 74 – 75, 156
Fluoride, effects of, 147 – 148
Flying
  as risk factor, 40
  risks of, 49, 156 – 157
Foege, W. H., 61
Food, adulteration of, 133 – 134
Food and Drug Act of 1906, 134
Food and Drug Administration
  food bannings by, 63
  labeling requirements of, 147 – 148, 169 – 170
Fossil fuels, risks from, 156
Free radicals, 112
Freedom, vs. risk management, 67
Freud, Sigmund, 93
Furocoumarins, 126

Gallbladder cancer, 142
Galton, Francis, 25, 32
  life of, 26
  work of, 26 – 28, 30 – 32
Gaming, as precursor to probability theory, 14 – 16
Gender
  risky behavior and, 101 – 102
  and sensation seeking, 101
Genes, 110
  and cancer, 114 – 115
Genetics
  and risk averseness, 89
  sampling in, 25 – 27
  traits determined by, 97 – 98
Geometry, analytical, 14
Georgia (former Soviet Union), longevity in, 60
Gibberellins, 126
Global warming, 8, 47, 156
God, speculations on believing in, 12, 15 – 16
Government, risk management by, 54 – 55, 63, 65 – 66
GRAS list, 137
Graunt, John, 19, 20 – 22, 32
Greeks, ancient
  personality theories of, 94
  probability studies of, 14
  technology of, 62
Green Revolution, 64
Grilled food, toxins in, 124
Growth hormones, 126

Hacking, Ian, 16
Harvard Center for Risk Analysis, 101
Hazard, origin of word, 17

Hazardous materials, 62
Health
panics regarding, 4
risk factors and, 3 – 4
Heart disease
incidence of, 50, 61
risk factors for, 31
Hemlock, 125
Herbicides, 139. *See also* Dioxin
Hexachlorophene, 140
High pressure nervous syndrome, 92
Hiroshima, 154
HIV
incidence of, 75
risk analysis of, 45
Hohenheim, Theophrastus. *See* Paracelsus
Holmes, Oliver Wendell, 5
Homeostasis, 80
risk, 80 – 82
Hormones, 98
artificial, 144 – 145
Horseradish, 126
Household accidents, 36 – 37
age and, 45 – 46
incidence of, 49
Human body
adaptability of, 106, 107 – 108
priorities of, 106
risk management by, 106 – 107, 108 – 109
Humphrey, Hubert, 117
Hunter-gatherer societies, 55 – 56
risk taking in, 103 – 104
Huygens, Christiaan, 18
Hydrazine, 126
Hydrogen peroxide, 112
Hydroxyl radicals, 112

ICMESA, 140
Industrial Revolution, chemical use in, 133
Information revolution, 3
Inoculation, 58
Insurance
birth of, 17
risk analysis in, 25
Intelligence, heritability of, 97
Iodine, toxic dose of, 132
Ionizing radiation, 112
risks from, 151

James, William, 3
Jansenists, 11, 15
Jefferson, Thomas, 174
Jenner, Edward, 57, 58, 59

Kelvin, Lord, 19

L-dopa, 99
Land development, effects of, 156
Langewiesche, William, 162
Law, and risk management, 65 – 66
Lead
as poison, 62
food adulteration with, 134
Legal system, overburdening of, 174
Leibniz, Gottfried Wilhelm, 17
Lichtenstein, Sarah, 72
Life expectancy, increase of, 60
Lifestyle, and life expectancy, 61
Linear, no-threshold model, 136, 137
Liver cancer, 142
Lloyd's, 17
Lois, Augusto, 114
London, demographics of, 21
Luther, Martin, 131

Manhattan Project, 154
McDonalds, 66
McGinnis, J. M., 61
Measles, incidence of, 59
Meat, carcinogenicity of, 151
Meat Inspection Act, 134
Meda, 140

Media
abuses by, 77 – 78
and public perception of risk, 73
Medici, Catherine de, 135
Medicine
risk management in, 167 – 169, 172, 174 – 175
use of probability in, 17
Mega-accident
as result of small errors, 158 – 160
risk of, 158
Merchant of Venice, 17
Mercuric sulfide, 134
Mercury
toxicity of, 133
universal occurrence of, 128
Méré, Chevalier de, 13
Microwaves, 112
Mitochondria, 112
Moivre, Abraham de, 23, 27, 32
Molecular biology, 108 – 109
Molecular epidemiology, 118 – 119
Monoamine oxidase (MAO), 100 – 101
Monoterpenes, 126
Mortality statistics, 50
Motorcycle helmets, benefits of, 41
Mouth cancer, risk factors for, 151
Mouthwash, carcinogenicity of, 151
mRNA, 110
Multiple myeloma, 142
Multivariate analysis, 102
Mutagens, 129

Nader, Ralph, 63
Nagasaki, 154
National Council on Radiation Protection and Measurements, 112
National Highway Traffic Safety Administration (NHTSA), 40, 41
National Resources Defense Council, 77
National Safety Council, 42, 44
National Transportation Safety Board, 157
Natural products, perceptions about, 76
Natural variation, study of, 25
Nero, emperor, 130 – 131
Neurotransmitters, 98 – 99
and risk taking, 100 – 101
Newton, Isaac, 23, 163
Nickoloff, Jack, 113
Nitrogen cycle, fertilizers and, 64
Nitropyrenes, 124
Norepinephrine, 98
Normal curve, 23 – 25
Novelty seeking, 99 – 100
Nuclear power
dangers of, 160 – 161
risks and benefits of, 170 – 171
Nuclear weapons, 154, 156
NutraSweet, 155

Obesity, risks from, 61, 151
Occupational Safety and Health Administration (OSHA), 63
Oldstone, Michael, 57
On the Origin of Species, 26
Opium, 126
Optimistic bias, 71 – 72
Ott, Wayne, 129
Outrage, 75
Ovarian cancer, risk factors for, 151
Oxalic acid, toxic dose of, 132
Oxidants, 113
Ozone layer, depletion of, 156

P-4000, banning of, 63
p53 gene, 115
and cancer, 116 – 119
Paciuolio, Luca di, 13
Paracelsus, 131
on toxins, 131 – 132, 143 – 144
Paradichlorobenzene, 128
Parkinson's disease, 99

Pascal, Blaise, 6 – 7, 32, 37, 46, 56, 164
  life of, 13, 15
  psychological profile of, 11 – 12
  studies on probability, 12 – 19
Pascal's triangle, 14
Pascal's Wager, 12
Pasteur, Louis, 59
Pasteurization, 59
Pavlov, Ivan, 94
Pearson, Karl, 25
Pedestrians, risks to, 47
Perchloroethylene, 128
Perrow, Charles, 157 – 158, 162
Personal injury lawsuits, and risk management, 65 – 66
Personal responsibility, 67
Personality
  ancient Greek theory of, 94
  factors affecting, 95 – 99
  Freudian approach to, 93
  inheritance and, 98
  twin studies of, 96
Pesticides
  benefits of, 146
  natural, 124 – 125
  regulation of, 135 – 136
Pitchblende, 153
Plague, bubonic, 57
  cause of, 21
  17th century epidemics of, 20 – 22
Plants, chemical defenses of, 126
Plutonium, 154
Pneumonia
  incidence of, 60
  treatments for, 59
Poisoning, 65
  incidence of, 49
Poisons, naturally occurring, 125 – 126
Polar bear metaphor, experiential basis for, ix–xii, 69 – 70
Polio, treatments for, 59
Polling, sampling in, 25
Pott, Percival, 133
Probability theory, 7
  birth of, 12 – 19, 56
  precursors to, 14 – 15
  uses of, 19 – 20
Problem of points, 13
Progressive movement, 134
Propyl mercaptan, 126
Proteins, 98
Proto oncogenes, 114
Prozac, 99
Psoralen, 126
Pyrethrins, 126
Pyrolysis, toxins generated by, 124

Quantitative toxicology, 127, 129

Radiation
  cellular effects of, 111 – 112, 113
  risks from, 151
Radioactivity, 76
  discovery of, 153
  fallout, 154
Radium, 129
  discovery of, 153
  effects of, 153 – 154
  uses of, 153
Radon, 76
Ramnefjell, 82 – 85
Recombinant technology, 63
Rectal cancer, risk factors for, 151
Regression coefficient, 30 – 31
Regression to the mean, 27 – 28
Relative risk, 19
Responsibility, personal, 67
Reversion to the mean. *See* Regression to the mean
Risk
  addiction to, 1
  alternative scenarios of, 47 – 50
  associated with low-probability events, 7 – 8
  assumptions about, 73

Risk (*continued*)
attacking head-on, 148–150
balancing, 80
balancing sources of, 148
calculation of, 38–39
of catastrophe, 158
definitions of, 6
environmental, 47
everyday, 35–39
everyday sources of, 3–4
group response to, 69–71
Hollywood view of, 157
ignorance about, 5
individual tolerances for, 82–83
information gathering about, 19–20
intuitive theories of, 71
language of, 8–9, 71, 79, 171, 175
mental models of, 71
modeling of, 7–8, 164–165
modern attitudes toward, 46–47
new types of, 4
opportunities provided by, 1
origin of word, 17, 18
perception of, 8, 39–40
physical reactions to, 90
propensity toward, 89
public perception of, 72–73, 75–77, 156
time dimension of, 46–47
traditional approaches to, 2–3, 4–5
unconscious preconceptions about, 71
unpredictable sources of, 2
voluntary vs. involuntary, 77
Risk analysis, 2, 17, 148
baseball analogy with, 43–44
behavior alteration through, 42–44, 50–51
birth of, 6–7
of carcinogenicity, 144
emergence of, 64
origins of, 18, 28
and risk management, 64–65, 171
Risk averseness, genetics and, 89
Risk communication, 72–73
Risk factors, 31
Risk management, 53–55
bodily, 105–106
computer use for, 164–165
consumer responsibility for, 168–176
decisions in, 148–150
for complex systems, 163–166
effects of, 55, 59–62
freedom and, 174
governmental, 54, 63–64, 65–66
historical origins of, 55–56
and personal freedom, 67
and personal responsibility, 67
origins of, 64–65
overuse of, 65–67
Risk taking
birth order and, 97
evolutionary benefits conferred by, 102–103
hormonal influence on, 99–101
in hunter-gatherer societies, 103
sensation seeking and, 95
youth and, 88
Roberts, John, 129
Rock climbing, 87–88
Romans, ancient
probability studies of, 14–15
technology of, 62
Roosevelt, Theodore, 2

Saccharin, 156
benefits of, 151–152
possible carcinogenicity of, 138
testing of, 143
Sacks, Oliver, 99
Safrole, 126
Salt, toxic dose of, 132
Sampling, 24
San Francisco earthquake of 1906, 164

Schwing, Richard, 41, 43
Scientific revolution, 56
Scrotal cancer, 133
Seat belt
  benefits of, 39, 41, 150
  costs of, 150
Sedentary lifestyle, risks from, 61
Sensation seeking, 95
  evolutionary benefits conferred by, 103
Serotonin, 98
Seveso, dioxin incident in, 140 – 142
Sex, risks from, 36
Shah, Hemant, 164, 165
Shakespeare, 17
Sidransky, David, 117
Silent Spring, 63
Sleepware
  regulation of, 54
  risks from, 156
Slovic, Paul, 72, 74 – 75
Smallpox, 22
  eradication of, 57 – 59
Smoking
  and cancer, 113, 116, 118, 120
  as risk factor, 39, 41, 61, 151
Spinach, toxic dose of, 132
Squamous cell cancer, 117
Streep, Meryl, 78
Stress, exhilaration from, 90
Stroke, 61
Sugar, toxic dose of, 132
Sulloway, Frank, 96 – 97
Sunburn, 116
Superfund sites, criteria for cleaning, 170
Superoxide, 112

Tacitus, 130
TCDD, 139, 140
Technology
  nonintuitive nature of, 72
  pervasiveness of, 155 – 156
  regulation of, 63
  risks posed by, 62
Teratogens, 129
Thalidomide, 143
Thomas, Lewis, 37
Three Mile Island, 160 – 161, 163
Titus, 130
Toothpaste, 147 – 148
Tort system, 174
Toxicology, 7
  algorithms of, 137
  future issues in, 145, 146
  quantitative, 127, 129
  recent advances in, 127 – 128
Toxicology (Casarett and Doull), 135
Toxins
  cumulative effect of, 129
  public perception of, 130
Tuberculosis, conquest of, 60
Tumor suppressor genes, 114 – 115
Twin studies
  of behavior, 96
  of novelty seeking, 100
Typhoid, conquest of, 59

Ultraviolet light, dangers of, 76, 111 – 112
Unemployment, risks from, 36
Union Carbide, 156
Unsafe at Any Speed, 63
Uranium, radioactivity of, 153
Utility maximization, 81

ValuJet Flight 592, 162
Vaughn, Diane, 162
Vegetables, toxins in, 124, 125 – 126, 148
Vermilion, 134
Vietnam, herbicide use in, 140
Vinci, Leonardo da, 131
Visibility, and perceived risk, 73
Vitamin D, toxic dose of, 132

Wald, Matthew, 137
Water, bottled
  perceptions about, 76
  toxic dose of, 132
Watson, James, 109, 100, 118
Weinstein, Neil, 71
Whiskey, toxic dose of, 132
Wilde, Gerald, 80 – 81
Wiley, Harvey, 134 – 135
William the Conqueror, 19
Wilson, Richard, 38 – 39, 45
Women, US life expectancy of, 60
World Health Organization, 58
World War I, 62

X-rays, 112
  discovery of, 152 – 153

Yersin, Alexandre, 21
Yogurt, possible carcinogenicity of, 151

Zuckerman, Marvin, 95, 100 – 101, 104